THE USE OF HUMAN BEINGS IN RESEARCH

PHILOSOPHY AND MEDICINE

Editors:

H. TRISTRAM ENGELHARDT, JR.
*Center for Ethics, Medicine, and Public Issues,
Baylor College of Medicine, Houston, Texas, U.S.A.*

STUART F. SPICKER
*School of Medicine, University of Connecticut Health Center,
Farmington, Connecticut, U.S.A.*

VOLUME 28

THE USE OF HUMAN BEINGS IN RESEARCH

With Special Reference to Clinical Trials

Edited by

STUART F. SPICKER

*School of Medicine,
University of Connecticut Health Center,
Farmington, Connecticut, U.S.A.*

ILAI ALON

*Faculty of Humanities, Tel Aviv University,
Tel Aviv, Israel*

ANDRE de VRIES

*Emeritus Faculty of Medicine, Tel Aviv University,
Tel Aviv, Israel*

and

H. TRISTRAM ENGELHARDT, JR.

*Center for Ethics, Medicine, and Public Issues,
Baylor College of Medicine,
Houston, Texas, U.S.A.*

KLUWER ACADEMIC PUBLISHERS

DORDRECHT / BOSTON / LONDON

Library of Congress Cataloging-in-Publication Data

The use of human beings in research.

(Philosophy and medicine; v. 28)
Includes bibliographies and index.
1. Human experimentation in medicine—Moral and ethical aspects. 2. Clinical trials—Moral and ethical aspects. I. Spicker, Stuart F., 1937– . II. Series.
R853.H8U84 1988   174′.28   88–6816
ISBN 1–55608–043–3

Published by Kluwer Academic Publishers,
P.O. Box 17, 3300 AA Dordrecht, The Netherlands.

Kluwer Academic Publishers incorporates the publishing programmes of D. Reidel,
Martinus Nyhoff, Dr W. Junk and MTP Press.

Sold and Distributed in the U.S.A. and Canada
by Kluwer Academic Publishers,
101 Philip Drive, Norwell, MA 02061, U.S.A.

In all other countries, sold and distributed
by Kluwer Academic Publishers Group,
P.O. Box 322, 3300 AH Dordrecht, The Netherlands

All Rights Reserved

© 1988 by Kluwer Academic Publishers, Dordrecht, The Netherlands
No part of the material protected by this copyright notice may be reproduced or
utilized in any form or by any means, electronic or mechanical including photocopying,
recording or by any information storage and retrieval system, without written
permission from the copyright owner

Printed in the Netherlands

TABLE OF CONTENTS

PREFACE / Robert U. Massey and Andre de Vries   vii

EDITORS' ACKNOWLEDGEMENTS   ix

STUART F. SPICKER / Introduction   xi

SECTION I / HUMAN EXPERIMENTATION AND THE LEGACY OF NUREMBERG

KENNETH L. VAUX AND STANLEY G. SCHADE / The Search for Universality in the Ethics of Human Research: Andrew C. Ivy, Henry K. Beecher, and the Legacy of Nuremberg   3

SECTION II / THE DEVELOPMENT IN MEDICINE OF THE IMPERATIVE TO CONDUCT RESEARCH WITH HUMAN SUBJECTS: AN HISTORICAL ANALYSIS

ROBERT U. MASSEY / Cultural Contents in the History of the Use of Human Subjects in Research   19

WILLIAM BYNUM / Reflections on the History of Human Experimentation   29

HANS-MARTIN SASS / Comparative Models and Goals for the Regulation of Human Research   47

CORINNA DELKESKAMP-HAYES / Moral Appropriateness in Human Research   91

AMOS SHAPIRA / Public Control over Biomedical Experiments Involving Human Beings: An Israeli Perspective   103

SECTION III / ETHICAL AND EPISTEMOLOGICAL ISSUES IN RANDOMIZED CLINICAL TRIALS

H. TRISTRAM ENGELHARDT, JR. / Diagnosing Well and Treating Prudently: Randomized Clinical Trials and the Problem of Knowing Truly — 123

STUART F. SPICKER / Research Risks, Randomization, and Risks to Research: Reflections on the Prudential Use of "Pilot" Trials — 143

ANNE M. FAGOT-LARGEAULT / Epistemological Presuppositions Involved in the Programs of Human Research — 161

MICHAEL RUSE / At What Level of Statistical Certainty Ought a Random Clinical Trial to be Interrupted? — 189

ANDRE de VRIES / Comment on Michael Ruse's Essay — 223

SECTION IV / OBLIGATIONS AND THE AVOIDANCE OF INJURY

ARTHUR L. CAPLAN / Is There an Obligation to Participate in Biomedical Research? — 229

T. FORCHT DAGI and LINDA RABINOWITZ DAGI / Physicians Experimenting on Themselves: Some Ethical and Philosophical Considerations — 249

IRVING LADIMER / Protection of Human Subjects: Remedies for Injury — 261

APPENDIX

ISRAEL HEALTH REGULATIONS: EXPERIMENTS ON HUMAN SUBJECTS – 1980 — 275

NOTES ON CONTRIBUTORS — 283

INDEX — 285

PREFACE

This volume, which has developed from the Fourteenth Trans-Disciplinary Symposium on Philosophy and Medicine, September 5–8, 1982, at Tel Aviv University, Israel, contains the contributions of a group of distinguished scholars who together examine the ethical issues raised by the advance of biomedical science and technology. We are, of course, still at the beginning of a revolution in our understanding of human biology; scientific medicine and clinical research are scarcely one hundred years old. Both the sciences and the technology of medicine until ten or fifteen years ago had the feeling of the 19th century about them; we sense that they belonged to an older time; that era is ending. The next twenty-five to fifty years of investigative work belong to neurobiology, genetics, and reproductive biology. The technologies of information processing and imaging will make diagnosis and treatment almost incomprehensible by my generation of physicians.

Our science and technology will become so powerful that we shall require all of the art and wisdom we can muster to be sure that they remain dedicated, as Francis Bacon hoped four centuries ago, "to the uses of life."

It is well that, as philosophers and physicians, we grapple with the issues now when they are relatively simple, and while the pace of change is relatively slow. We require a strategy for the future; that strategy must be worked out by scientists, philosophers, physicians, lawyers, theologians, and, I should like to add, artists and poets.

This work, which has undergone extensive revision since the days of the symposium in Israel, examines a few of the issues in that strategic plan, and compares analyses on several points of view that will enable us to grow a little in our own understanding so that we can be more useful to those who must shape health policies for the future.

Farmington, Connecticut, U.S.A. ROBERT U. MASSEY

Allow me in this part of the Preface to recall an ancient clinical trial described in the *Old Testament*. I cite the text as given in the *New English Bible* (Book of Daniel I, 1–16):

PREFACE

In the third year of the reign of Jehoiakim, king of Judah, Nebuchadnezzar, king of Babylon, came to Jerusalem and laid siege to it. The Lord delivered Jehoiakim, king of Judah, into his power . . . and he carried them off to the land of Shinar. . . . Then the king ordered Ashpenaz, his chief eunuch, to take certain of the Israelite exiles, of blood royal and of the nobility, who were to be young men of good looks and body without fault, at home in all branches of knowledge, well-informed, intelligent, and fit for service in the royal court; and he was to instruct them in the literature and language of the Chaldeans. The king assigned them a daily allowance of food and wine from the royal table. . . . Their training was to last for three years, and at the end of that time they would enter the royal service. Among them there were certain young men from Judah called Daniel, Hananiah, Mishael and Azariah.

Now Daniel determined not to contaminate himself by touching the food and wine assigned to him by the king, and he begged the master of the eunuchs not to make him do so. God made the master show kindness and goodwill to Daniel, and he said to him, "I am afraid of my lord the king; he has assigned you your food and drink, and if he sees you looking dejected, unlike the other young men of your own age, it will cost me my head." Then Daniel said to the guard whom the master of the eunuchs had put in charge of Hananiah, Mishael, Azariah and himself, "Submit to us this test for ten days. Give us only vegetables to eat and water to drink; then compare our looks with those of the young men who have lived on the food assigned to them by the king, and be guided in your treatment of us by what you see." The guard listened to what they said and tested them for ten days. At the end of ten days they looked healthier and were better nourished than all the young men who had lived on the food assigned them by the king. So the guard took away the assignment of food and the wine they were to drink, and gave them only the vegetables. . . .

Now this clinical trial, successful as it was, lacked several of today's customary criteria – the project was executed in spite of its rejection by the authority; there was apparently no informed consent by the control group; the study was neither randomized nor double blind; and the methodology of evaluation was, to say the least, rather crude.

In all probability Daniel's research proposal would be rejected today by any review committee – it would be judged scientifically and ethically non-valid. Still, Daniel's clinical trial was canonized in *The Bible*, and is read over and over again throughout the world. A perusal of the documents concerning our ancient past, then, reveals a heritage that has bearing on the general theme of this volume. It is worthwhile to keep this in mind as the reader turns to the principal contributors of this volume and their careful philosophical, medical, legal, and religious reflections on the use of human beings in research.

Tel Aviv, Israel   ANDRE de VRIES

EDITORS' ACKNOWLEDGEMENTS

The Editors take this occasion to acknowledge their gratitude to TEVA Pharmaceutical Industries Ltd., Tel-Aviv, Israel, Tel Aviv University and its Faculties of Humanities, Law, and the Sackler School of Medicine. Appreciation is also extended to the co-hosts of the symposium, The Tassia and Dr. Josef Meyshan, Chair of History and Philosophy of Medicine at Tel Aviv University Medical School, and the Division of Humanistic Studies in Medicine of the School of Medicine, University of Connecticut Health Center, Farmington, for providing not only the material support that made possible the convening of the Fourteenth Trans-Disciplinary Symposium on Philosophy and Medicine at Tel Aviv University – but for creating a convivial and hospitable atmosphere that made the visit to Israel a joyous experience for all participants and guests. A special thanks is extended to Yoram Dinstein, J.D., the previous Rector of Tel Aviv University, and Robert U. Massey, M.D., previous Dean of the School of Medicine, University of Connecticut, for their many kindnesses, support, and co-hospitality. The Editors wish also to offer their mutual appreciation to Theo Ungewitter of the Department of Community Medicine and Health Care, who inherited the final secretarial tasks to complete the manuscripts, and Sarah Beckett, J.D., of Farmington, Connecticut, who volunteered to read through each page of the manuscript in order to refine the texts and prepare the Introduction. We are indebted to Susan G.M. Engelhardt, the Assistant Editor of the series, for kindly and unselfishly agreeing to correct the final manuscript and page proofs, and to Lorraine Langlois for preparing the Index in order to include the volume in the *Philosophy and Medicine* series.

INTRODUCTION

It is commonplace for the public to associate admission to hospital with a rapid transformation from an active, functioning life – with all of its social demands – to the more passive role of patient. Notwithstanding the recent movement toward patient autonomy where he is as active as possible in his own care, still the role of patient is sharply distinguished from one's everyday roles. Perhaps this shift is strongly influenced by the existential condition of illness, infirmity, disability, or even some very vague malady not easily captured by the nosology of contemporary clinical medicine. Whatever the explanation, it is clear that most patients do not anticipate that they may become the object of an educational enterprise that has been structured to tutor students of medicine, nursing and the allied health professions. The role of *patient*, however, particularly in a teaching hospital, may give way to the role of *educational subject*, with the two oscillating throughout the hospital stay. Quite often, moreover, a third role is suggested: the *research subject*.

For it is often overlooked that many persons are frequently asked to participate in a study, research program, or scientific protocol, in order to expand "the frontiers of knowledge" ([2], p. 1), as the President's Commission has expressed it in its report of 1981, *Protecting Human Subjects*. It appears that a consensus has evolved which maintains that research, especially research which includes the participation of human beings, is integral to medicine as a science, as well as medicine at the bedside. Indeed, there is a growing consensus that clinical acumen integrated with the methods of experimental inquiry can obtain valid knowledge, in a scientifically rigorous sense, which is acquired in the clinical context itself; that is to say, the bedside becomes the laboratory. But it is clear to everyone that at all stages of life the participation of human beings in the conceptual designs known as experiments may be a potentially risky enterprise. The United States and many other nations, therefore, have reached a consensus that, notwithstanding the high value placed on the prevention and cure of illness, relief from pain and suffering, and the postponement of death, research conducted by persons with expertise ought not to be carried out on competent others without first taking numerous precautions, even if the participants have provided their informed, voluntary consent. In short, no *unnecessary* risks should be suggested by researchers. Indeed, this principle became quite focal in the President's Commission's report, which is prone to reflect the consensus if not always sound argument.

If one traces the history of experimentation with human subjects – which at times includes autoexperimentation by researchers – one finds that the United States is not alone in formally registering its concern over the welfare of persons in their role as research subjects. Many other nations have extensive research programs that are founded on ethical principles and enforced by regulation, even if the entire scientific edifice is not yet on stable ground, e.g., we have not yet achieved an equitable and just system of compensation for persons injured as a direct result of their participation in research protocols. But a system that adequately provides for equitable compensation is perhaps not too far off. Many insurance models could begin to fill the space.

In order to open a dialogue among scientists, philosophers, theologians, attorneys, and other interested persons, a group of American and European scholars visited Israel in September, 1982. Israel, of course, has had a significant research enterprise since its inception, and yet discussions between Israeli professionals, who had dedicated their lives to expand the frontiers of knowledge, had scarcely taken place in a context that focussed on the ethical aspects of conducting research with human subjects. Still, there was a common bond, and in a certain sense a common history. That common past, reflected in the transcripts of the *Tokyo War Crimes Trials* and especially the [*Nuremberg*] *Trials of War Criminals* [4], serves to connect, in a very special way, Israel's sensitivity to the fate of human beings who voluntarily participate in the search for new knowledge that can contribute to the alleviation of pain and suffering or, for that matter, any of the goals of humane medical practice.

We in the United States are also strongly affected by the memory of innumerable "scientific" atrocities, but we have tended to learn more vicariously and cannot therefore pretend to a wisdom we have not earned.

To reflect a commitment to the values that would permit only the conduct of humane research enterprises, the volume opens with a co-authored essay by a theologian and a physician. Their commemoration of Andrew C. Ivy, M.D., and Henry K. Beecher, M.D., serves to admonish us, without preaching emptily, that we must never rest in comfort and lazily presume that all is well in the world, when human experimentation proceeds with the coerced participation of tens of thousands of persons. To be reminded of the dangers of the past is, unfortunately, too often shrugged off; in fact it is more than prudential that we remain alert to the ever-present need for raising conceptual and moral concerns in the context of research with human subjects. One only has to recall, as H. Tristram Engelhardt, Jr., has pointed out in *The Foundations of Bioethics*, that not only does the Tuskegee syphilis study, which was supported by the U.S. Public Health Service, demonstrate "the need for the special protection of human subjects, one would need to add that the mere publication of codes and laws does not appear to be sufficient to protect human subjects." As Hans-Martin Sass has shown, Engelhardt adds, "quite exacting and in many respects progressive rules for the protection of human subjects were in fact enacted in Germany in 1931 and were theoretically in force until 1945" ([1], p. 291; also see [3]).

Section I, "Human Experimentation and the Legacy of Nuremberg," contains a recent historical overview of the ethical issues in research with human subjects. Kenneth Vaux, a theologian, and Stanley Schade, a physician, critically review Dr. Ivy's work on the Nuremberg Code, and the later efforts of Dr. Henry K. Beecher to formulate an ethical agenda generated from the Nazis' extensive research with human subjects. Ivy's and Beecher's discussions of scientific excellence, humanitarian law and principles, and case-study illustrations lead to their analysis of the meaning of altruistic care, a specifically theological notion. Hence, they conclude that both "philosophical and mystical perfectionism are to be found in the formulation of the Nuremberg Code." By introducing the remaining essays from the perspective of the aberrant and abhorrent "experiments" articulated at the Nuremberg trials, the authors define the volume's orientation.

Robert U. Massey, M.D., opens Section II, "The Development in Medicine of the Imperative to Conduct Research with Human Subjects: An Historical Analysis," by describing the crucial and powerful role of medical science today, and the falling away of the 19th-century ethos regarding science and medicine. Future advances are likely to be so rapid, and by implication of such force and power, singly and collectively, that to deal with them ethically will require all our energy and the adoption of new strategies by physicians and non-physicians alike. In the U.S., funds for research have been large, and have increased four hundredfold during the past twenty years. So much is new that treatment that is, in fact, virtually experimental soon becomes commonplace. Medical research has become so vast a competitive endeavor, therefore, that it raises incrementally the potential for moral danger. Fortunately, we have been reasonably prepared because this scientific movement follows a prior commitment to humanitarian ideals. Massey also notes the antipathy towards science, which, along with other trends, has already shaped the expression of experimentation with human subjects and other aspects of research, as well. He reminds us that there is no turning back, since biomedical research has become the chief intellectual work on which clinical medicine now rests.

William Bynum, an historian of medicine, underscores the threads of commonality among the various types of medicine, which have existed from the earliest historic times for which we have records. He takes us from the Hippocratic, largely observational medicine of ancient Greece, through the period of the Roman Empire, and even beyond. He argues that there has always been human experimentation in the context of the practice of medicine, especially experiments upon the economically disadvantaged. Medicine, however, was vastly different from its present form, and it took a slow evolution of theory and institutions (and concomitant fluctuations in the credibility of medical practitioners) before the advent of actively experimental medicine in the late nineteenth century. The establishment of scientifically based clinical research, along with the mutual dependence of clinical service and experimental medicine, led to the rapid scientific advance of medicine, and the formulation of specifically ethical problems. These problems have come to characterize the context for modern investigators, who, at the same time, maintain that experimentation is at the very heart of modern medicine.

Hans-Martin Sass, a philosopher, assesses critically the current model used to assure the safety and security of subjects who participate in research protocols – that is, the relatively recent *regulatory model*. He notes that the earliest (and possibly most stringent) code was developed in Germany in 1931, and that later the Nuremberg Code set out the requirements of today's system – among them the skill required of the researcher, the need for positive risk-benefit determination, the voluntary consent of the subject, and the requirement of independent review boards. Sass's criticism centers around the fact that regulation is presumed to serve as a substitute for the moral responsibility of the physician-researcher, and the fact that all types of risk in human research (even the most unexpected or minor risks) are treated equally. Sass points out that controlled clinical trials may actually be enormously wasteful, duplicative, and even unethical, if they seriously delay treatment (as some have recently complained about delays in federally approving the prescribing of pharmaceutical substances that may prevent or cure AIDS). Finally, Sass worries that the present scheme (as it exists, for instance, in West Germany) may undercut personal responsibility and thus diffuse legal liability.

Sass's proposals are offered to correct this situation by substantial deregulation and by establishing a system to compensate properly subjects who participate in biomedical research (along with the shouldering of personal responsibility by researchers). He even suggests the possibility of drawing more persons, including professional subjects, into the pool of potential subjects. Workers' compensation insurance, which is provided by commercial insurers on a no-fault basis, would go far toward providing reliable recompense. This could replace present over-stringent testing, which often precedes the approval of new pharmaceuticals, as well as the continuous assessment of such drugs. This, in time, would yield a heightened awareness on the part of all concerned, and thus stress professional and moral responsibility on the part of physician-researchers.

Corinna Delkeskamp-Hayes, a philosopher, proposes a novel, time-and-bureaucracy-saving form of regulation that is based on the autonomous response of subjects, but not *solely* on their sometimes quite elusive informed consent. The concept that she introduces is that of *endorsability* under which all research subjects would comment on the conduct of experiments and how they as subjects have been affected; the results of their comments – even if only about their anxiety, discomfort, or pain – would be included in the evaluation of the researcher, but in a formally institutionalized way. Trust between physician-researcher and subject would remain intact: the patient could continue to trust the physician-researcher to conduct and administer his treatments, but afterwards the manner of the treatment would be subject to formal and careful review. In the middle-risk range of human experimentation, such comments from participating research subjects could be used to endorse or not to endorse the researcher's methods.

Recognizing that there would be some serious resistance to her proposal, Delkeskamp-Hayes outlines the ways such resistance might be overcome in the domain of ethics, political theory, and policy-making.

In an imperfect world this model of compromise – leaving only some aspects of control and power with the morally responsible physician-researcher, and some with the public body who participated in the research protocol – may tend to be a better solution, and other theoretical objections would thus be overcome more easily.

Amos Shapira, a professor of law, argues that the issues surrounding biomedical research on human subjects are very complex, and that inadequate attention has been focused on reinforcing the decision-making power of the individual *vis-à-vis* powerful, paternalistic medical institutions. Shapira recognizes and appreciates the powerful attraction of the model of self-supervision by the experimenter, but admonishes us that today the issues are too complex to be left solely to the judgment of researchers. The models of private, institutional control *vs.* public control are assessed by Shapira with specific reference to Israel's concerns. Shapira analyzes in detail the present situation in Israel, with respect to experimentation with human subjects, and what legal institutions and laws are currently available to regulate such research. In this regard he notes with approval the emergence of "Helsinki Committees" and public health regulations, which, in 1980, represented the first legislative and supervisory structure over the domain of human experimentation in Israel. According to Shapira, much remains to be done in terms of non-procedural, concrete formulations of public policy where novel, challenging problems arise daily; but much is being done in terms of rule-making and continuous trans-disciplinary discussions.

H. T. Engelhardt, Jr., introduces Section III, "Ethical and Epistemological Issues in Randomized Clinical Trials," arguing that since all empirical knowledge is provisional, *bona fide* experimentation ought to be treated as *prima facie* for the good (leaving room for the well-founded caution that there is always the possibility of abuse and exploitation of the poor and defenseless). There is a human compulsion to treat diseases and to seek to postpone death. The question of experimentation arises in the context of an institution already committed to intervene in all illness. Due to the probabilistic nature of knowledge claims, there will always be risks to the individuals who participate in knowledge acquisition, but these risks are not necessarily incompatible with their wishes and aims and their own personal cost-benefit calculations. The physician-researcher and patient may, therefore, comfortably respect each other's position and still proceed with randomized clinical trials (RCTs) and double-blind studies in the great majority of cases.

Stuart Spicker, a philosopher, explores the ethics of conducting RCTs in the context of the ethical responsibilities of the physician to protect his patient. Given that some patients in an RCT will not receive a new treatment that is believed to be more efficacious than what they currently receive, when should one conduct a pilot study which simply tests the effect of the new treatment and deprives no one? Spicker argues that pilot trials, not RCTs, conducted with careful observation need not be biased or unethical: randomization in the strict statistician's sense is not always required or necessary from the very outset, if one seeks sound experimental results; his view he places in contrast with that of Dr. Chalmers.

Furthermore, there is no real danger when an RCT is conducted, that it will be undertaken with too few subjects to present a statistically valid result; this bias is due to widespread misunderstanding of significance levels in experimentation. In an important study discussed by Spicker, where only a small fraction of the published results of various studies were obtained with sufficient subjects to ensure significant results, validity was not compromised. If this is true, there is even more reason to consider pilot studies as an alternative to RCTs.

Anne Fagot-Largeault then applies the microscope to the standards employed in biomedical research on human subjects, both those professed and those practiced. Her thesis is that during the past several centuries there has been an almost imperceptible shift away from the absolutism of medical beneficence, to a new relativism that enables research to be performed on human subjects. Formerly, she points out, the benefits of research were thought to accrue largely to individual researchers though also to patients; today, the decision to do research is "society's decision," with all the potential for coercion which that term already implies. She demonstrates that society's choice is often rooted in ambiguity. That is, scientific credos apparently presume the following: (1) knowledge is innocent; (2) seeking new knowledge is inevitable; (3) science, by applying "good" methodology, will ensure truth; (4) "minimal-risk" experiments may be easily defined; (5) physicians may be trusted to understand statistical methods and appropriately to limit their experiments, given the end in view; (6) the results of research may rightly be applied on a broad scale (e.g., a new vaccination), even when unknown side effects are likely to occur. Without in any way condemning clinical or experimental science, Fagot-Largeault uncovers some critical presuppositions that often lead the unwary researcher into error.

Michael Ruse examines, from the point of view of a philosopher, at what point an experiment involving risk to human subjects ought to be terminated. First, he explores the concepts of (1) significance, (2) errors involving accepting or rejecting the null hypothesis, and (3) 1-tailed and 2-tailed experiments. These concepts, familiar to the statistician, are quite necessary to Ruse's subsequent analysis of the interests of the various participants in conducting an experimental trial: the researchers, the subjects, those who suffer from an illness, the physicians who are not researchers, the sponsors of the trial, and the public. He concludes that while values extrinsic to the experiment itself must enter in, we should beware of overly stringent criteria for permitting human subjects to engage in a biomedical experiment. Circumstances as well as the availability of other information about a pharmaceutical substance may require the researcher to modify what would otherwise be standard precautions. He cites instances in which drug testing in the U.S. often delayed the entry of medicinals already in use in Great Britain to the almost certain detriment or even early demise of some patients. Thus, high significance levels and powerful test designs may sometimes have to yield to the immediate exigencies of active treatment. The problem is to determine when this is justified.

Andre de Vries comments on what he takes to be the central thesis of Ruse's essay: that there may be good reasons not to stipulate or require very high significance levels in drug testing; at times, it may be reasonable to accept far-from-definitive or inadequate evidence. The real question for de Vries is: Who is to determine the appropriate level of stringency for a specific trial? Such a determination involves not only sensitive questions of adequate evidence, but raises economic and moral issues as well. Should the physician decide? or, perhaps, the patient-subject? The practical complexities of this problem surely bristle; they are left implicit in Ruse's essay and thus require further reflection, according to De Vries.

T. Forcht and Linda R. Dagi, two physicians, introduce Section IV, "Obligations and the Avoidance of Injury," by exploring one aspect of the "romantic paradigm of medicine" about which there has been little comment: autoexperimentation. In experimenting on himself, the physician acts at once in two roles: healer and potential victim. The physician may have a better opportunity, acting in this dual role, for acute observation without harming any subjects. To the element of inquiry may be added another component, which we praise: suffering for the common good. Suffering, of course, may impair his work (or even kill him).

We should, however, according to the authors, have reservations about physician autoexperimentation. For example, Jewish law forbids hazardous self-experiments unless one is in desperate need of therapy. However, what is seen as wrong by society may be permissible or even morally required when the reason for the physician's assuming a real risk is that he wishes others to assume the same risk with him. The willingness of the physician to undergo autoexperimentation then becomes a laudatory or even requisite precondition for involving other human agents.

Of course, experiments that involve only one person and no control are subject to objection if certain types of statistical validation are being sought, whether or not the experiment poses significant risks. In other situations, though, particularly when subjective insights are sought, the auto-experiment may be morally justified.

In any event, the authors go on to argue that there is no obligation on the part of the physician to subject himself or herself to life-threatening experiments. The physician, in evaluating the benefit of the experiment, must be as honest with himself about himself and his personal expectations in conducting the experiment as he would be with others who consent to participate in the experiment at his request.

Arthur L. Caplan, a philosopher, argues that most of the attention by those who study the many facets of human experimentation have been directed to protecting those who consent to participate, not to assuring that there are statistically adequate numbers of persons to participate. There may well be inequities in the way research is performed, for in large part it is performed on the sick, institutionalized, or the indigent. Caplan argues that it is morally reprehensible for the rest of society to escape its share of the obligation to participate, and thus take a "free ride."

Caplan then argues that there exists such an obligation; it is properly grounded not in "social contract" theory or a theory of achieving a hypothetical "public good," but rather on the notion of "fair play" – that is, if one intends to seek out the benefits of modern research, one is obliged to offer to participate in that research. Finally, Caplan argues that the state should have no part in enforcing this moral duty; also, one who chooses to opt out of a particular institution that expects participation in research should certainly be permitted to do so. While it may not be easy to identify all persons who benefit from various forms of biomedical research, certainly many more should be included in this population.

Irving Ladimer, an attorney, argues for a formal system to compensate those who participate in biomedical research to be determined by personal injury and documented as causally related to the research, but without determination of fault. His proposals, though long-standing, are yet to be accepted. The preference of those empowered to establish such a system seems so far to be opposed to establishing one, and thus rests on existing theories of negligence and/or *ad hoc* remedies. Arguably, the numbers of persons involved in biomedical and behavioral research and entitled to some compensation has become a moral issue, even if all documented, serious injuries are few in number. On the basis of fairness, then, as well as simplicity, Ladimer argues for a system similar to a worker's compensation program, in which injuries incurred on the job are compensated for on the basis of a predetermined monetary schedule. Significantly, other options, like arbitration or litigation, are not necessarily foreclosed; given efforts toward finding a solution based on a formal and coherent policy, rather than on merely informal remedies that reflect the present situation, it may be that private interests will, in time, reach a viable solution to no-fault injury, without imposing governmental sanctions or creating additional bureaucratic mechanisms.

The volume closes with an Appendix – Israel's Health Regulations, approved in 1980 – which reflects the principles that govern all research performed on Israeli soil, or sponsored by Israeli agencies and conducted by Israeli scientists. In fact, Amos Shapira's essay frequently signals the importance of these Regulations and reflects the legal and ethical principles upon which all Israeli research with human subjects must be conducted. By applying the principles of the Health Regulations, it follows (1) that not all experimentation with human subjects is morally condoned and, (2) that individual rights, both legal and moral, may even serve to restrict certain goals (e.g., the acquisition of new medical knowledge) as well as what may clearly be judged a social good.

School of Medicine STUART F. SPICKER
University of Connecticut Health Center
Farmington, Connecticut, U.S.A.

BIBLIOGRAPHY

1. Engelhardt, H. T., Jr.: 1986, *The Foundations of Bioethics*, Oxford University Press, New York.
2. President's Commission for the Study of Ethical Problems in Medicine and Biomedical and Behavioral Research: 1981, *Protecting Human Subjects*, U.S. Government Printing Office, Washington, D.C.
3. Sass, H.-M.: 1983, 'Reichsrundschreiben 1931: Pre-Nuremberg German Regulations Concerning New Therapy and Human Experimentation', *Journal of Medicine and Philosophy* **8**, 99–111.
4. *Trials of War Criminals Before the Nuremberg Military Tribunals: The Medical Case*: 1948, Vols. 1 and 2, U.S. Government Printing Office, Washington, D.C.

SECTION I

HUMAN EXPERIMENTATION AND THE LEGACY OF NUREMBERG

KENNETH L. VAUX AND STANLEY G. SCHADE

THE SEARCH FOR UNIVERSALITY IN THE ETHICS OF HUMAN RESEARCH: ANDREW C. IVY, HENRY K. BEECHER, AND THE LEGACY OF NUREMBERG

I. INTRODUCTION

The first ethical requirement of human studies is that of scientific excellence. Meaningful truth must be sought. This means three things: a worthy goal, sound design, and competent execution. Beyond scientific excellence, studies must possess two additional ingredients to meet the ethical requirement. We will call these the humanistic and altruistic imperatives. Humanistic law requires that we do not deliberately harm another. This is the negative dimension of ethics: non-maleficence. These evaluative perspectives built on the sure foundation of good science should be found whenever clinical investigations are undertaken in universities or hospitals. The members of institutional review boards, representing various disciplines, should assure that this tripartite examination is made of every protocol. Good science, protective and affirmative ethics should always be present. The assumption of this paper is that science itself is a value. Therefore, only excellent science can be ethical and only ethical science is of value.

Society today has taken up with increased intensity the perennial concern to make science ethical and humane. Society at large and the scientific community share this objective. This paper will critically assess the contribution of Andrew Ivy and the Nuremberg Code and the later contribution of Henry Beecher to our search for those timeless and universal ethics that should guide human research. Ivy, Vice-President of The University of Illinois Professional Colleges, was called in 1946 to be the principal medical consultant at the second Nuremberg trial, the trial of the physicians. He was a provocative and controversial figure. Ivy's testimony at Nuremberg reveals the perfectionism and paranoia of his complex personality. The enduring genius and problematic of the Code is the byproduct of these stigmata. Just as Ivy would safeguard the University from communism in the McCarthy Era and defend Krebiozen as the panacea for cancer, so with mingled suspicion and supererogation his beliefs would stamp the Nuremberg creed.

Ivy's testimony became the architecture for the Nuremberg Code, the

S. F. Spicker, I. Alon, A. de Vries, and H. T. Engelhardt (eds.)
The Use of Human Beings in Research. 3–16.
© 1988 *by Kluwer Academic Publishers.*

derivative Geneva and Helsinki formulations, and the ethicolegal standards that undergird clinical investigations to this day. The *Belmont Report* [2] and the *Guidelines for the Protection of Health and Human Services* [9] specifically build on this foundation. The disclosures and testimony of that trial set in motion an earnest endeavor forever after to establish human experimentation on the three moral pillars of excellent science, humanistic law, and altruistic care.

The troubling experience of the generations touched by the war still animates that search. The idealism of the Nuremberg Code was the fruit of Andrew Ivy's fear of evil and his yearning for physical *shalom* (well-being). This idealism in the Code, as Beecher and others have noted, is its besetting flaw; however, that idealism has also become the Code's most lasting distinction, arguing, as it does, the necessity for a positive moral disposition on the part of the patient and of the investigator, in order to augment protections phrased as prohibitions.

Let us examine the three requirements for ethical experimentation that are explicit and implicit in the Nuremberg Code, arguing that satisfactory biomedical protection exists only when these three are present and function in concert.

II. SCIENTIFIC EXCELLENCE

Scientific excellence is the *sine qua non* of ethical human experimentation. Scientific excellence means that the study is truthful (not deceptive) and significant (not trivial). The Code expresses this point in several ways:

Stipulate #2: The experiment should be such as to yield fruitful results for the good of society.

Stipulate #6: The degree of risk to be taken should never exceed that determined by the humanitarian importance of the problem to be solved by the experiment.

Stipulate #8: The experiment should be conducted only by scientifically qualified persons ([8], p. 288).

When knowledge becomes possible there is the moral imperative to pursue it. There is a moral and even a theological impulse to discover. In an early article Ivy wrote:

In the medical sciences, the only method which can clearly reveal and establish the cause, prevention, and treatment of disease is the method of controlled experimentation on animals and volunteer human subjects ([3], p. 2).

Determining whether an experiment represents good science is frequently far from easy. Facile appeal to the Stipulates may in fact impede good medicine.

For instance, debate now rages over the necessity of rigorous scientific study of therapeutics. Should double-blind, controlled studies be initiated even with medications and treatments already widely accepted and used? Some argue that we should undertake controlled studies even on such widely used substances as digoxin and coumadin. Others argue that ethics prohibit putting anyone on control arm or placebo when dealing with an apparently efficacious agent. These clinicians, arguing often on grounds of freedom of the investigator, say that we should continue our present practice of relying on historical and experiential wisdom. In refutation, by contrast Dr. Arnold Relman, editor of *The New England Journal of Medicine*, contends that many of our treatments, medicinal and surgical, are untried. In the use of digoxin for congestive heart failure and in the technique of coronary bypass surgery, cesarean section delivery or 5-azidothimidine (AZT) for AIDS, for example, it remains unclear whether the efforts are efficacious, benign, or harmful. The question of when and with whom they are efficacious remains unsettled. H. Tristram Engelhardt goes beyond this to argue that the continuing scandal of untested medical therapies is not only scientifically irrational but unconscionable because of the injury to those we presently maltreat or fail to treat.[1] To these persons and those in the future who might stand to benefit, we have the moral obligation to do rigorous clinical science.

Also, we need to remember that an insight is available only when its time has come and the human mind is ready to know it. Much scientific work fails to display obvious immediate benefit. It establishes base-line data from which the scientist can work toward breaking through to new knowledge and techniques. Moreover, surprise and serendipity are constant landmarks in the history of science. An investigator who feels he or she is on to something, or is building a block of knowledge that will ground another discovery, should be encouraged.

We often evaluate scientific merit by review panels in relation to the funding source. Professional jealousy is omnipresent, as is professional arrogance. Einstein cautioned that valuable discovery is often unexpected. It is important to retain some flexibility in determining what indeed is important.

Some feel that Ivy himself made the error of being unwilling to accept the slow-plodding accumulation of knowledge in cancer research that had, in fact, characterized his own early physiological investigations. Wanting a breakthrough in cancer treatment, or even it is said (perhaps unfairly) a Nobel prize, Ivy was tarred by the Krebiozen scandal.

Nevertheless, Ivy made a crucial contribution at Nuremberg. He restated the historic commitment to good science, which is found from ancient times in medicine down to the dawn of modern work with Claude Bernard, but which commitment was so notably absent in the Nazi experiments, which included the "science" of killing (ktenology). Studies on humans must be preceded by a sound conceptual design, *in vitro* studies, bench studies, and finally animal studies. The careful, scientifically scrupulous accounts of Ivy became the informing criteria for the Nuremberg Code as scientific excellence was established as the primary ethical requirement of biomedicine.

III. HUMANITARIAN LAW

To meet the ethical requirement, science must be excellent. But good science does not fulfill the ethical equation. To ensure that moral quality is present in scientific work it is necessary to move beyond science to the realm of value. Excellent science can be made captive to malignant political purpose. Return to the scene of the Nuremberg trial.

Karl Brandt and the other German physicians and scientists were asked if their actions conformed to the law. "What law?" they asked, "We were obeying the laws, executing the policies and following the orders of the government." They were correct. By this time the sterilization, racial hygiene, and euthanasia programs had been enacted. Absurdly, the State had also issued an edict prohibiting animal experimentation.

The Nuremberg Tribunal began to search for a higher, universal law that transcended national laws. They began to interrogate Dr. Ivy, seeking to establish what they called the universal laws of humanity.

When Ivy had been invited to be the medical and ethical expert at Nuremberg, he immediately assembled all the great books of moral philosophy from Aristotle to Kant, and the legal and constitutional texts explicating the jurisprudential foundations of human rights and freedoms. He studied them carefully and shared with the court, as well as any person trained in science could, the humanistic foundations of Western philosophy and law as these applied to medical experimentation.

The judges asked the question concerning universal law in two ways: First, what has happened where civilization endures; second, historically, how has universal law evolved? Western culture in both the Graeco-Roman and Judeo-Christian traditions has believed in natural law, a moral substance that is imbedded in God, nature, and history. This law is palpable and knowable just as are the laws of physical nature rooted in the empirical world. Are there universal, cross-cultural ideals and values that do not vary with time and space? Are there imperatives that are rooted in the genetic and biological process? How shall this universal moral substance be designated? The idea has been refined across history.

There is a moral law writ in the fibre of nature. This was the claim of all the Western-Oriental cultures as conscience dawned when the last Ice Age withdrew and civilized communities formed. The Sumerians, Babylonians, and Egyptians proclaimed with the Habiru and the noachite law, the primal universal covenant that binds all mankind, life with life. It is a completely negative law. It states the *prohibitiva*, what must not be done. In this code, blasphemy, sexual misdirection, harm to fellow persons and animals were proscribed. Anthropologically speaking, these were the temptations that human flesh was heir to. This universal natural law which Thomas Aquinas called *lex naturalis*, as opposed to *lex divina*, was to be found in the architecture of the universe itself.

Kant and the great Enlightenment philosophers agreed with Plato, Seneca, and the Ancients that the moral imperative resided in the very structure of reason, or at least at the boundaries of reason. For Kant, reason acquaints us with the moral principles that are *a priori* to reason itself. The common sense of men is compelled by reason itself to posit the existence of a moral universe and the ethical imperative.

Through his testimony, Ivy spelled out the fundamental rights: access to information, freedom of one's person from violation, consent, and the opportunity to withdraw. They were all articulated in detail as they apply to medical investigations.

Ivy related how human studies were undertaken in the United States, specifically in the case of those studies with which he was familiar. In the Naval Research Institute, for example, studies with aims similar to Nazi "experiments" had taken place: researching the desalinization of sea water, the effects of high altitude, and the treatment of malaria. These studies were going on in the Department of War, the universities in

Chicago, and in Statesville Prison. That the Tuskegee (Alabama) syphilis studies occurred at the same time as the Nazi studies is a sobering memory.

Ivy outlines the manner in which volunteers were solicited, how consent was obtained from them, and how disengagement from the study was facilitated. In a series of studies under the auspices of the State of Illinois, with subjects in state institutions, volunteers were solicited in a particular way: notices were placed, risks clearly outlined, "rewards" eliminated in coercive settings. According to Ivy, volunteers had to be turned away, despite all the warnings and disclosures. Truly informed consent, and the right to withdraw at any time, he felt, could and should be a reality. This formidable requirement is embodied in the first stipulate of the Nuremberg Code:

1. The voluntary consent of the human subject is absolutely essential. This means that the person involved should have legal capacity to give consent; should be so situated as to be able to exercise free power of choice, without the intervention of any element of force, fraud, deceit, duress, overreaching, or other ulterior form of constraint or coercion; and should have sufficient knowledge and comprehension of the elements of the subject matter involved as to enable him to make an understanding and enlightened decision. This latter element requires that before the acceptance of an affirmative decision by the experimental subject there should be made known to him the nature, duration, and purpose of the experiment; the method and means by which it is to be conducted; all inconveniences and hazards reasonably to be expected; and the effects upon his health or person which may possibly come from his participation in the experiments.

The duty and responsibility for ascertaining the quality of the consent rests upon each individual who initiates, directs or engages in the experiment. It is a personal duty and responsibility which may not be delegated to another with impunity ([8], p. 227).

Just twenty years after Ivy testified before the Commission, Dr. Henry Beecher, writing in *The New England Journal of Medicine*, made a trenchant point about the need for a check upon the use of human subjects in a context completely divorced from Nazi Germany [1]. His point underlines the usefulness of universal criteria, which can be referred to and used in our era, particularly concerning the need for informed consent from human subjects.

Let us first restate the problem of using human subjects in the context of natural science: Since the Enlightenment, an important human value has been objective, scientific truth. To quote the Enlightenment philosopher, Immanuel Kant:

When Galileo experimented with balls of a definite weight on the inclined plane, when Toricello caused the air to sustain a weight which he had calculated beforehand to be equal to that of a definite column of water, . . . a light broke upon all natural philosophers. They learned that reason only perceives that which it produces after its own design; that it must not be content to follow, as it were, in the leading strings of nature, but must proceed in advance with principles of judgement according to unvarying laws and to compel nature to reply to its questions. For accidental observations made according to no preconceived plan cannot be united under necessary law, but it is this that reason seeks for and requires. . . . Reason must approach nature with a view, indeed, of receiving information . . . not in the character of a pupil who listens to all his master chooses to tell him, but in that of a judge who compels the witnesses to reply to those questions which he himself thinks fit to propose. To this single idea must a revolution be ascribed by which, after groping in the dark for so many centuries, natural science was at length conducted into the path of certain progress ([4], p. 6).

This statement embodies the human value of desiring to know the truth and to know reality. But what of the conflict when reason seeks to understand the mechanism of growth of a human cancer cell, or seeks to compel nature to give the answer to the question: What are the best treatments for Hodgkins' Disease, acute leukemia, or sickle cell anemia? The problem is that the disease comes attached to a human being, a rational being. When Kant wrote of the natural physical sciences, he could extol the scientist as compelling nature to answer reason's own questions according to reason's own designs. But when the same philosopher wrote of human morality, he brought forward another principle, that of the autonomy and freedom of every rational being not merely as a means to an end, but always as an end in himself.

Natural scientists may design an experiment in human disease so that an answer is compelled from nature, but any such experiment must disregard the autonomy of the rational being who bears the disease. The investigator can, of course, in experiments on animals compel nature at a biological level below that of rational beings. But ultimately, the scientist desires to compel an answer from the disease state as it exists in the rational being – man.

The resolution of this conflict has been attempted by the doctrine of informed consent. According to this doctrine, the rational being consents to the experiment of his own free will, making the purposes of the experiment his own ends and thereby becomes no longer simply a means to an end. The process of informed consent is fraught with many obvious difficulties. The technical nature of the experiment may render it difficult to communicate to the subject, or the patient-doctor relationship may compel agreement to an experiment, eliminating the subject's freedom and autonomy. Fear of the disease itself may rob the subject of his autonomy and his free will. A further problem is that the rational being carrying the disease may not be well served by hearing the exact nature of his disease in every detail. We realize that in these times it is a sacrosanct principle that patients be told every detail of their disease. Nevertheless, there are often times when patients cry, "Enough." Psychoanalysts may spend hours listening to their patients before being certain that they can tell the patient the results of their analysis without causing offense or upsetting the patient. Physicians treating cancer or other chronic and lethal diseases may find themselves in the position of delivering the most terrible news to individuals after only a few minutes of casual psychological observation.

A further problem is the recent finding that a patient will choose surgery for lung cancer over radiation therapy when "outcomes are framed in terms of the probability of survival rather than in terms of the probability of death. . . . This effect of using different terminology to describe outcome represents a cognitive illusion" ([7], p. 1260). We believe that this effect, wherein decisions are pre-structured to elicit inconsistent responses, will further reduce our confidence in our ability to obtain rational, informed consent.

We are tempted to ask, "Does it really all matter?" Can we not leave in the hands of the physician the design of experiments that will yield what we desire: objective truth with minimum risk to the subjects? Can we avoid proliferation of forms, committees, and informed consent documents? We have little evidence to answer these questions. As members of institutional review boards (IRBs), we often refer to events in Nazi Germany as a reason for the development of the IRB system, and the emphasis on informed consent. We sometimes sense among investigators that they feel there is no real need for such a system, and that medical experimenters could perfectly well police themselves without outside review. To such investigators the Nazi example does not appear completely relevant.

The authors believe that Dr. Beecher, in 1966, in *The New England Journal of Medicine* marshaled the most pertinent data in examining the question as to whether investigators can and should regulate themselves [1]. He reviewed numerous articles that were published prior to 1966, and ascertained whether the information reported in those articles could reasonably have been obtained with the informed consent of the subjects involved. His guiding principle was that no patient or subject would agree to an experiment in which there was risk of harm to himself, without any benefit to himself. He reviewed 100 articles in a leading medical journal and concluded that 12 of the experiments described were unethical, that is, that no reasonable individual could have agreed to participate in them since the risk of the experiments to the patients was so great. From another set of articles that he reviewed, he reported on 22 experiments he considered unethical.

It is instructive to reconsider some of the experiments that he discussed. For instance, it was known that penicillin usually prevented the non-suppurative complications of streptococcal respiratory infections. Despite this, definitive treatment was withheld and placebos were given to a group of 109 men in the military service, while benzathine penicillin was given to the others. This therapy was given according to the military serial numbers assigned to the subjects. In the group of subjects who did not receive penicillin there were two cases of acute rheumatic fever and one of acute nephritis; these complications did not occur in the men given penicillin.

A second study dealt with the question of whether sulfonamides were effective in shortening the duration of acute streptococcal pharyngitis and reducing its suppurative complications. Again, this study was carried on despite the knowledge that penicillin was effective in preventing the development of rheumatic fever. The subjects were a large group of hospital patients. The control group received no specific therapy, the test group received sulfadiazine, and in the control group 4.2% developed rheumatic fever. Theoretically, the use of penicillin could have reduced these suppurative complications to less than 1%.

In a third study, the treatment of typhoid fever with chloramphenicol was investigated. It was known at that time that chloramphenicol was an effective treatment for this disease. Despite this, the study was undertaken to determine the relapse rate under the two methods of treatment. Of 408 charity patients, 251 were treated with chloramphenicol. In this group 8% died. Symptomatic treatment was given but chloramphenicol was withheld in another 157 of whom 36 (23%) died. According to the data presented, 23 patients died in the course of this study who would not have been expected to succumb if they had received specific therapy.

A fourth study described children undergoing cardiac surgery for congenital heart disease. They were randomized into two groups. In the first group a total thymectomy was done. In both groups full thickness skin homografts from an unrelated adult donor were sutured to the chest wall. The study was done as part of a long range study of the growth and development of these children and to examine the effect of thymectomy on immunologic resistance in children. At that time it was known from experiments on mice that early thymectomy would result in immunologic paralysis of the adult animals. Dr. Beecher believed it extremely unlikely that parents would have consented to such studies on their children.

In another study the antibiotic triacetyloleandomycan was given to juveniles residing in a detention home. It was suspected that this antibiotic would cause liver disease. The subjects received the antibiotic for the treatment of acne, a weak indication. A large percentage did develop liver disease, and of those many had liver biopsies.

In still another case, a malignant melanoma was transplanted from a daughter to her volunteering and informed mother in the hope of gaining a better understanding of cancer immunity and in the hope that the production of tumor antibodies might be helpful in the treatment of the cancer patient. The daughter died on the day following the transplantation of the tumor into her mother, so the hope seems to have been more theoretical than practical. The implant in the mother was excised on the 24th day after surgery. She died from metastatic melanoma on the 451st day after transplantation. Evidence that she had died of diffuse melanoma that metastasized from a small piece of transplanted tumor was considered conclusive.

Dr. Beecher also discussed the experiments in which cancer cells were implanted into elderly patients without their informed consent that they would be given subcutaneous injections of tumor cells. These studies were done to investigate the effect of immunity in the development of cancer ([1], p. 358).

When one recounts details of such experiments, which were performed during the 1960's in the United States, the reaction of most lay audiences can best be described as one of astonishment verging on horror. We can recall reading some of these articles before the publication of Dr. Beecher's essay, and can remember discussing some of them at journal clubs. In general, the ethical dimension of these articles, which was highlighted by Dr. Beecher, was not a serious subject taken up for discussion. One might consider that some investigators were overzealous or perhaps lacked the proper degree of caution. However, the real focus was on whether the proper scientific question had been put to nature, and whether the proper answer had been compelled from her by the design of the experiment. That is, did the experiment represent good science. Following the publication of Dr. Beecher's article, an insight came to many medical investigators. The problem of consent had to be addressed, among others, and it was no longer possible in the United States to discuss the ethics of human experimentation in the abstract as though it were unrelated to problems of experimentation and design. These problems were not limited to aberrant states of the past, but also clearly belonged to the present.

Investigators at that time did argue that the studies reported by Beecher were exceptional and not typical of most clinical experimentation. They argued that effective control of human experimentation should reside only with the individual investigator and any limitation on the investigator would result in stagnation in the progress of medical science. Dr. Beecher himself argued for control of unethical experiments as the responsibility of the editors of medical journals. He asked that editors reject for publication descriptions of experiments they deemed unethical. In effect, he argued that data obtained without attention to the ethical dimension should be regarded as invalid.

The authors believe that we should re-examine in a positive way the validity of Dr. Beecher's original thesis. Recently, there has been much discussion of the problem of fabrication of data in medical research. This issue appears to have aroused even more concern within the medical research community than the issue of honesty with regard to informed consent. These issues are related: Honesty in informed consent is as important as honesty in the production and publication of experimental results.

At present, honesty in the analysis and production of data is self-controlled; administrative controls are not imposed to ensure honest data. Researchers should have the same respect for honesty in informed consent, and we should have little need for administrative control of this consent.

The lesson of Dr. Beecher's paper was the need for investigators to acknowledge and be aware of the ethical dimension of their work. The ethical dimension is intrinsic to any human experiment, and it should in principle require as little outside administrative control as the analytic dimensions of the experiment.

We realize that this argument could be driven in the other direction; that is, that more administrative control is needed for the maintenance of honesty in data reproduction itself. We confess that we do not know how to disarm that argument. We do, however, wish to see honesty and openness in human scientific investigation with minimal administrative control, but a high degree of internal self-control. We think that a periodic re-reading of Dr. Beecher's essay can contribute to that end.

IV. ALTRUISTIC CARE

Finally, however, medicine like religion involves a deeper ethical covenant than prudence or jurisprudence can fathom and which neither human reason nor the law can fully implement. Therefore, at the horizon of ethics we find the altruistic imperative. Altruistic care invites more than obligation demands. It is what the Hebrew comes to know as *hesed* and the primitive Christian community comes to call *agape*. Beneficence or the more powerful *caritas* or mercy is a disposition more of the heart and the will than the mind. It is the disinterested concern for the well-being of another.

In the final hours of Ivy's testimony at Nuremberg, he was asked what he thought it took to keep the practice of medicine moral. Can the state and the law create and sustain the moral atmosphere necessary for the right conduct of medicine and human investigations? "No," Ivy replied, "the state cannot follow a physician around . . . to see that moral responsibility is properly carried out." The moral responsibility of the Oath of Hippocrates must be inculcated into the mind of every physician. "It is the Golden Rule of medicine. . . . It states how one would wish to be treated by another if he were ill . . . that is the way a doctor should treat his patient or experimental subject. . . . The moral imperative of the Oath of Hippocrates is necessary for the survival of the scientific and technical philosophy of medicine" [3].

Appeal to the Golden Rule and to the Hippocratic Oath is at once an appeal to enlightened self-interest and to the perfectionist ideals of philosophy and religion. The Golden Rule in one version is rooted in the old Hammurabic *lex talionis*, "You get what you give." The Hippocratic Oath can be seen as a document of physician aggrandizement. At another level, however, these great affirmations aspire to the highest ethical perfectionism. There are two traditions of moral perfectionism that guide Western ethics: the philosophical tradition expressed in Kant and the other idealists, and the Christian ethical tradition embodied in the Sermon on the Mount ("Be Ye perfect as your heavenly father is perfect") [6]. Kant argues that the ontological concept of perfection is a supreme rational principle of morality. It draws moral decision away from the level of "sensibility" to "the court of pure reason" ([5], p. 60). It was the biblical understanding of moral perfectionism, however, instilled by his Wesleyan heritage that shaped the moral convictions of Andrew Ivy. Wesley taught that the moral life entailed both purity and discipline. To do right in personal action meant to be chaste in thought and deed and lawful in personal interactions.

Both philosophical and mystical perfectionism are to be found in the formation of the Nuremberg Code. Some say this renders it unrealistic and irrelevant. The section on informed consent, for example, is so thoroughgoing and demanding that few clinicians feel it can be honored in every detail. In principle, however, it is an excellent standard. But, as Beecher pointed out, "Rule 1 states a requirement very often impossible of fulfillment" ([8], p. 232). In situations of medical duress, powerlessness, dependence and fear, self-assertion fades and one becomes a victim at worst – at best, a case.

In this situation the virtue of mercy must be added to those of humanitarian law and scientific integrity. But mercy is not care that is compulsive and legalistic, but care that bears a profound sense of cohumanity and empathy.

Medical science today is a treacherous yet urgently required endeavor. More and more will medicine rest risky inquiries. Heart, lung, and liver transplantation and cancer therapeutics, where the power to heal becomes the power to kill, are just two examples. The Nuremberg Code, thanks to the beliefs of Andrew Ivy and the work of men like Henry Beecher, promulgates values worthy of taking us into a new age.

University of Illinois
Chicago, Illinois, U.S.A.

NOTE

[1] Personal conversations with Drs. Relman and Engelhardt.

BIBLIOGRAPHY

1. Beecher, H. K.: 1966, 'Ethics and Clinical Research', *New England Journal of Medicine* **274**, 1354–1360.
2. National Commission for the Protection of Human Subjects of Biomedical and Behavioral Research: 1978, *The Belmont Report: Ethical Principles for the Protection of Human Subjects of Research*, US Government Printing Office, Dept. of Health,

Education and Welfare (now Health and Human Services), Washington, DC, publication # (OS) 78–0012. Appendix 1–0013, Appendix 2–004.
3. Ivy, A.: 1948, 'The History and Ethics of the Use of Human Subjects in Medical Experiments', *Science* **108**, 1–5.
4. Kant, I.: 1787, 'The Critique of Pure Reason', preface to the Second Edition, Vol. 42, *Great Books of Western World*, (ed.) Robert Maynard Hutchins, p. 6, 1952.
5. Kant, I: 1949, *Fundamental Principles of the Metaphysics of Morals*, Bobbs-Merrill, New York, pp. 59–60.
6. *Matthew* **5** : 48.
7. McNeil, R. G., Parker, S. G., Sox, H. C., and Tuersky, A.: 1982, 'On the Elicitation of Preferences for Alternative Therapies', *New England Journal of Medicine* **306**, 1259–1261.
8. *Nuremberg Code*, cited in Beecher, H.: 1970, *Research and the Individual*, Little, Brown & Co., Boston, pp. 227–232.
9. *Protection of Human Subjects*, 45 Code of Federal Regulations, 46, Office of Protection from Research Risks Reports, March 8, 1983.

SECTION II

THE DEVELOPMENT IN MEDICINE OF THE IMPERATIVE TO CONDUCT RESEARCH WITH HUMAN SUBJECTS: AN HISTORICAL ANALYSIS

ROBERT U. MASSEY

CULTURAL CONTENTS IN THE HISTORY OF THE USE OF HUMAN SUBJECTS IN RESEARCH

What if, at some future date, we were to come across an account of this congress on 'The Use of Human Beings in Research,' and discovered that it had been held in Joppa in 1882? We should believe almost surely that we had been made the victims of some historical prank; it would be as though we had found a manuscript of a musical composition identical to Beethoven's String Quartet in B flat major with a date of 1725 rather than 1825, and, more than that, had found that the composer had lived in Jerusalem rather than in Vienna. Scientific, philosophical, legal, and religious works belong to particular times and places, are part of a *Gesamtkultur*; we have a feeling about where they fit, in whose company, in which decade, in what place. Science, art, literature, religion, and politics, which make up the parts of an entire culture, are recognizably members of the same household.

Professor Bynum will shortly take us through the successive historical stages of 2500 years of Western medicine and will show how the cultural, professional, epistemological, and technical setting of each stage has affected attitudes toward sickness and health and has determined the work of physicians, their education, and how they have acquired their knowledge and skills. Without knowing the historical background we can make but little sense of our own foreground of 1982, with its peculiar set of cultural, professional, epistemological, and technical characteristics. In order to know where we are, we must first find out where we have been.

Historical continuities may appear at times to be broken, especially when other cultures force their way in, or when there are social and scientific revolutions; however, even in periods of rapid change there are probably no true discontinuities.

Forty years ago little was written about the ethical issues involved in doing research with human subjects; twenty years later we were appointing human research review committees and asking whether informed consent was possible for most of the individuals who might agree to be experimental subjects. What happened during those two decades? Certainly the Nazi horror and the revelations at the Nuremberg trials made us sensitive to the possibilities for great evil in using human beings as objects of experimental work.

But further, in the United States federal expenditures for research in medical schools had increased 400-fold during those 20 years, from $780,000 to $307,000,000 ([8], p. 198). In the 1930s and 1940s the term 'informed consent' had not been invented, clinical research was by and large an individual and informal activity, subjects were volunteers who more often than not had a trusting personal relationship with the clinical investigators as patients, students, or colleagues.

What have been the changes in attitudes and values since the beginnings of experimental medicine and clinical science over a century ago; more especially what has occurred in the past 40 years which has led to these ethical concerns in 1982 and which makes this symposium so timely?

As our knowledge and our power grows, our judgment about what we are to do and what we ought to do is tested in new and unfamiliar ways. Some ethical problems may be recognized only within their historical setting, and are not definable outside of it. Further, the recognition that there are ethical problems in human experimentation is not a sign of our generation's more refined and highly developed ethical sensitivity. Sir Thomas Browne's testimony at a trial at Bury St. Edmunds in 1669 of two women accused of witchcraft may have contributed to their conviction and hanging; his expression of belief in witches derived from his religious faith and would have found support among most of his colleagues and contemporaries; in fact, to deny the devil and his angels in the mid-seventeenth century would have given rise to suspicions of atheism, and would have been viewed by some important people as a dangerous opinion threatening to morality, law, and the social order. Sir Thomas Browne's ethical sensitivity and standards of morality may have been neither higher nor lower than ours; he was a man of seventeenth century Europe who had written a book entitled *Vulgar Errors*; yet he believed that the king's touch was effective against scrofula. Few of us move far from our own time and place.

All medical practice from earliest antiquity has involved physicians trying things out on their patients. Given the uncertainties of diagnosis, prognosis, and therapy, physicians could do little more than frame hypotheses about the nature and prognosis of their patients' illnesses; therapy was a way to test those hypotheses.

But that kind of speculating with nature is hardly the same as research, at least not the same as modern research in the stage of laboratory or experimental medicine.

There had indeed been ethical concerns in the old medicine before modern science. Galen, referring to the Hippocratic aphorism that declared experience to be dangerous (*experimentum periculosum*), wrote, "In the medical art, unlike other arts, the material is not brick, clay, wood, stone, earthen ware, or hides – . But in the human body to try out what has not been tested is not without peril, in case a bad experiment leads to the extinction of the whole organism" ([4], p. 4).

These are the same perilous or fallacious experiences that make up the everyday professional life of practicing physicians. Bernard Towers wrote of modern medical practice: "Most diagnostic and therapeutic techniques today are so sophisticated and 'scientific' that almost every other case becomes in fact an experiment in the traditional sense" ([10], p. 167).

Modern medicine rests on modern biological science, but practicing medicine is not the same as doing science. Often medical teachers tell their students that there is no essential difference between what a clinician does in the hospital and what a biomedical scientist does in his laboratory. That is not quite so. To be sure, clinical research and the care of patients may come together in the teaching hospital, and the clinical investigator at times may be unable or unwilling to distinguish his role as physician from his role as clinical scientist. This leads, however, to difficulties for the auditor who reviews research accounts and for the hospital administrator, as well as for the patient, the physician, and, if he is also a teacher, his students. Sorting out the rights and obligations of each person in this confusion of activities is no easy task of ethicists, lawyers, and research funding agencies.

Research has become a career for teachers of medicine, a major activity for teaching hospitals and academic medical centers, and essential for nations seeking to establish their worth, to maintain their prestige, their favorable balance of trade, or their strong military position. That is a far different endeavor from the work of a busy seventeenth-century practitioner like William Harvey pursuing his idea of the motion of the heart and the circulation of blood, or even a Howard Florey and his team in 1941 beginning clinical trials with penicillin at Oxford. Before 1940, clinical research was no more than an avocation for most members of the clinical faculty; clinical scholarship was more descriptive than experimental; and careers did not hang on successful competition for grant support.

Only thirty years ago, Professor William Beveridge wrote: "Until recent times research was carried on only by the devotees, because the material rewards were so poor, but nowadays research has become a regular profession" ([1], p. 203).

You may recall that in 1967 federal support for research in American medical schools was about $300,000,000; by 1979–80 that had increased to more than $1,000,000,000 ([7], p. 2927). Since that increase occurred during a period of rapid inflation, the real growth of the research enterprise may be better appreciated by noting that between 1950 and 1977 American biomedical research manpower increased from 15,000 doctoral scientists to 90,000, of whom 36,000 were physicians ([3], p. 162).

There are over 50,000 full-time faculty members in American medical schools. They, along with faculty members in most academic medical centers in the Western world, are unlikely to enjoy a distinguished career, or even to hold their positions and earn their livelihood, unless they are successful in competing for research grants and having their work published in refereed journals. Medical schools depend on their research productivity to attract outstanding faculty, to enroll capable students, and to compete for financial support from government agencies and private donors. Research has taken on many of the characteristics of an industry, and has recruited experts in management, financial administration, production, science writing, and marketing, along with a host of supporting personnel, research assistants, and graduate students. The goals of research today include much more than the advancement of knowledge or the relief of human suffering. Research is no longer the hobby of a devoted few but rather a career for thousands, the product of great research institutes, universities and their academic medical centers, and an essential work for industry and commerce.

Research teams are assembled, strategies are developed, supporting groups organized for marketing and for developing capital, and special societies formed for political action. Governments recognize that industries that advance biotechnology are essential to national prestige and to the economy. A recent article in *Science* comments: "The dominant postwar philosophy in British research has been that the principle criterion for supporting science should be excellence as judged by fellow scientists. In contrast, calls are increasingly being heard from both government and industry that support for basic research should be more focused on areas of potential social or industrial need" ([2], p. 813). Western society, and indeed all of mankind, is just beginning its latest voyages of discovery.

The magnitude of the biomedical research endeavor, the numbers of men and women whose livelihood depends on it, the careers made or broken by success or failure in research, the commercial, political, and national interests at stake have raised questions over the matter of conflicting interests and the potential for moral danger. These issues simply were not raised by the work of clinical scientists 100 or even 50 years ago. Yet the ethical questions are neither unique nor new for our time. Louis Pasteur faced these concerns when he made the decision to use his untested rabies vaccine on the young boy, Joseph Meister, 120 years ago. Robert Koch faced them when he tried *tuberculin*, the glycerine filtrate of cultures of the tubercle bacillus, as a treatment for tuberculosis. Do the potential benefits outweigh the risks? Equally important, does the experiment or the clinical trial represent good science? The question that was always hard to answer was whether the subjects or patients understood what was being done to them; were they informed when they gave their assent? And because they had consented, and had become volunteers, did that make everything all right?

In the days of Koch and Pasteur the public worried more over animal experimentation, where informed consent was not possible, than over human experimentation, where consent could be freely given. Antivivisectionism remains a matter of serious ethical concern, but whether consent can be sufficiently informed to be voluntary now seems less clear than it did in the 19th century.

Our attitude toward the use of human subjects in clinical research may have changed, along with the growth of research and the development of scientific medicine, for other reasons which may not be unrelated to that development. The idea of a democratic society, ideas about universal human rights, the humanitarian movement, and the growth of science are all manifestations of the cultural evolution of our society since the Enlightenment that are without parallel in history. For the 50 years following the French Revolution there were rapid changes in social and political thought in Europe and America. They were accompanied by the brilliant rise of scientific medicine, especially in Paris, during the first half of the 19th century. The philosophical empiricism of the French *Idealogues* and the mechanism of Newton found an antiphonal response in the *Naturphilosophie* of the German romantic movement, the first favoring clinical science, the second generally antithetical to it. Between the ideas of individual rights and human liberty coming from the Enlightenment and expressed in political revolution west of the Rhine, and the idealism and moral zeal of romanticism east of the Rhine, there developed a new set of attitudes toward human welfare that can only be called humanitarian. Richard Shryock notes that, "It may have been pure coincidence that the 'discovery' of anesthesia was concomitant with such other ameliorative developments as the movements to forbid flogging, or to provide a more merciful treatment of the insane." ([9], p. 175). It may not have been coincidence that the movement to abolish Black slavery began in those years, along with increasing sensitivity to the misery of poverty and the pain of chronic disease. Humanitarianism was a new impulse in art, science, literature, politics, and religion; the idea of medicine expanded to include the public health.

As the great advances in the physical sciences began to be matched by those in geology and the biological sciences, the romantic reaction against science and social change grew more intense. The popular fear of speculating with nature was certainly nothing new, but it was hardly diminished by the publication of Darwin's *Origin of Species* in 1859. The conflict between the evolutionary idea in modern biological science and popular religious tradition seems likely to continue into the 21st century. More serious is the perceived conflict between humanism and science which divides intellectuals today no less surely than it did 180 years ago. In the minds of many men and women in this age of technology and science, science represents a kind of dangerous or double-faced power for good and evil that will achieve mastery over the soul and the society of man, a product of the human mind over which mankind no longer can exercise control. In a recent survey in the United States, the National Science Foundation reported that more than one-third of those questioned believed that scientific discoveries "make our lives change too quickly" and "break down our ideas of right and wrong." Even among thoughtful and well-educated men and women, science and technology have been seen as being, on balance, if not sinister, at least inimical to morality and harmful to the natural environment. Research is considered to be an activity owned by the scientific community, in no way subject to control or guidance, and technology is perceived to be advancing inexorably by its own laws, quite out of reach of the law and out of touch with the purposes of the social order.

Even in medicine, in which science and technology have been seen as mostly benign, some claim to see as much mischief as good. The idea that medicine may have caused as much suffering as it has relieved is not limited to vegetarians and Christian Scientists. Alternatives to traditional medical care seem to appear as frequently today as they did in the days of bleeding and purging. The application of science and technology to the diagnosis and treatment of disease and the attitudes that an emphasis on science seems to engender in students and practitioners have been viewed as somehow antithetical to the caring aspect of the physician's work. Bernard Towers writes: "It inevitably follows that, for the moment, the student (and his teacher) sees the patient less and less as a person, as more or less integrated and whole, and more and more as a somewhat fragmental collection of thousands of variables. . . . Alienation, at the human level, is then invariable" ([10], p. 165).

This year we celebrate the one hundredth anniversary of Koch's demonstration that the tubercle bacillus is the cause of tuberculosis. Since 1882 the death rate from tuberculosis has fallen in most Western countries from 200 or 300 per 100,000 to less than one per 100,000. Life expectancy has increased by 25 years and is still rising. Even so, confidence in medical care is declining and attitudes of anti-science and anti-technology abound. It is a difficult time for clinical research, a time for scientists to work to restore popular confidence in science and medicine.

My intention has been to note three cultural currents in the history of the past two hundred years that may have contributed to our present attitudes toward clinical science and toward the use of human subjects in research. Experimentation on human beings has been a necessary part of medicine from the days of Hippocrates, and in the past 100 years clinical science and biomedical research have become the chief intellectual work of medicine. Attitudes toward that work have changed remarkably, especially in the past 40 years. Those attitudes, I believe, have resulted from the rapid growth of the research enterprise in the past century, especially since the Second World War; from the emergence and continued progress of the humanitarian movement; and from the antipathy toward science and technology which has been expressed in both popular and intellectual circles since the age of romanticism.

Bioethics is a new word I cannot find in my dictionary; as a word it may be something of a barbarism; its roots suggest that it should mean the ethics of life; we understand that it is the ethics of medicine and of the needs, rights, and obligations of patients, physicians and other health professionals, subjects, and clinical investigators. It has come into being with the growth of clinical research from a part-time occupation of busy medical practitioners, to a career for academic physicians, to the major activity of great institutes and academic medical centers. When we add to this the commercial interests involved in pharmaceuticals and medical technology, and the requirements for human research in those activities, it is evident that clinical investigation and biomedical research are no longer the avocation of a few curious physicians. It is not unreasonable to suppose that scientific interests, career concerns, institutional pride, and commercial considerations may not always coincide with the best interests of the human subjects of clinical research. Yet the value to society of this research work and the nature of man the experimenter make the continuation of science and technology a certainty. It is hard to imagine that we shall become bored with our science or turn away from our inventions.

"Man is an inveterate experimenter," wrote A. V. Hill in 1929.

Those of us who have been small boys ourselves, or indeed are still small boys, will know what joy is found in taking an old alarm clock to bits or a bicycle to pieces, in seeing how fast we can run a hundred yards, in breeding rabbits, pigeons, or canaries, in fixing wireless apparatus together; or when we get older, in trying a new kind of oil or petrol or even a new medicine [6].

Later in the same essay, he asks:

How, then, shall we proceed? By experiments on man? Surely, for man is the only experimental animal who will cooperate fully with the experimenter, and most important results have been obtained by controlled experiments on man.

Just as we have continued for 200 years to work out the political and social implications of the Bill of Rights, so we shall work out the implications of the humanitarian movement for medicine, for medical research, and for health policies.

And, finally, through education and a clearer public understanding of science and technology as human and humane enterprises, we may narrow the gap between the two cultures, put aside the fear that science must have about it something of the diabolical, and assert, with Bacon, that we seek to understand nature "for the uses of life."

Some years ago, Thomas Hunter concluded a presentation he had made at a conference on clinical research with a quotation from an editorial in *Science* by Willard Gaylin and Samuel Gorovitz. I should like to repeat it:

There is no turning back from science and technology. Man is driven by his nature to modify the conditions of his existence. He will not return by choice to early death and unnecessary suffering . But if science is to flourish, it must enjoy public understanding. It must make its case to those who are unconvinced, either because they are not aware of the issues or because they are not yet satisfied with the arguments they have heard. The attitude of paternalism which is characteristic of the doctor-patient relationship may be acceptable or even inevitable in a clinical setting, but it is wholly inappropriate in institutional settings and statements about scientific research.

Scientists and their critics must not merely state their positions, they must come to understand each other's point of view ([5], p. 315).

School of Medicine
University of Connecticut Health Center
Farmington, Connecticut, U.S.A.

BIBLIOGRAPHY

1. Beveridge, W. I. B.: 1950, *The Art of Scientific Investigation*, Vintage Books, New York.
2. Dickson, D.: 1982, 'British Universities in Turmoil', *Science* **217**(4562), 811–813.
3. Frederickson, D.: 1977, 'Health and the Search for Knowledge', *Daedalus* **106**, 159–170.
4. Galen: in *Hippocratis Aphorismos Comentarii I*, quoted by Temkin, 1975, in *Respect for Life, Medicine, Philosophy, and the Law*, Johns Hopkins University Press, Baltimore, Maryland.
5. Gaylin, W. and Gorovitz, S.: 1975, 'Academy Forum: Science and Its Critics', *Science* **188**, 315.
6. Hill, A. V.: 1960, 'Experiments on Frogs and Men', in *The Ethical Dilemma of Science*, the Rockefeller Institute Press and the Oxford University Press, New York, pp. 24–38.
7. Petersen, E. S. *et al.*: 1981, 'Undergraduate Medical Education', *Journal of the American Medical Association* **246**(5), 2913–2930.
8. Powers, L. *et al.*: 1967, 'Medical School Expenditures', *Journal of the American Medical Association* **202**(8), 758–762.
9. Shryock, R. H.: 1947, *The Development of Modern Medicine*, Alfred A. Knopf, New York.
10. Towers, B.: 1971, 'The Influence of Medical Technology on Medical Services', in G. McLachlan and T. McKeown (eds.), *Medical History and Medical Care*, Oxford University Press, London.

WILLIAM BYNUM

REFLECTIONS ON THE HISTORY OF HUMAN EXPERIMENTATION

I. INTRODUCTION

Public awareness of the scientific, ethical, legal, and social dimensions of medical research on human beings has increased dramatically since World War II. The Nuremberg Trials horrified the world by revealing the extent to which Nazi physicians had tortured, mutilated, and killed human beings in the name of medical science and the exigencies of war. Developments in medicine since World War II have done little to assuage concern, for along with the advances in understanding, diagnosing, treating, and preventing disease have come a battery of uncomfortable and potentially dangerous invasive medical techniques: more elaborate surgery, powerful drugs with often powerful side-effects, and the potential for behavior modification to an extent unparalleled in earlier history. As a result of much research utilizing human subjects, however, a more technically proficient medicine has been achieved. The growth of medical knowledge has had the paradoxical effect of enhancing the possibility that research-minded doctors may feel the pull between immediate patient care and more precise knowledge, or in other words, between their short-term responsibilities and their long-term desires to improve medical knowledge, thereby making more effective diagnosis and treatment available to future patients.[1]

These and related ethical dilemmas have confronted many doctors and lay people during the past few decades. From inside and outside the medical profession there have been calls for more open discussions and stricter controls on research on human subjects.

A change in attitude may be gleaned from two books of the same title published 22 years apart. Kenneth Mellanby's *Human Guinea Pigs* [31] recounted research work which he, a distinguished medical scientist, directed during World War II on diet, the transmission of scabies, and other topics. His volunteers were conscientious objectors who were prepared to contribute to medical research but not directly to Britain's war effort. In what was still effectively the pre-Nuremberg era, Mellanby seemed relatively unconcerned with ethical issues. He did not hide the fact that some of the research involved placing his subjects in potentially uncomfortable, inconvenient, or embarrassing situations, such as sleeping next to a person infected with scabies. Rather, he believed that the War had provided concrete examples of the value of using human subjects in experimental investigations and hoped that the return of peace would not mean the end of volunteers for human experimentation projects. So long as there was obligatory military service, he felt that young men would still participate as subjects in medical research as an alternative to military duty.[2]

By 1967, when Dr. M. H. Pappworth published his own book entitled *Human Guinea Pigs: Experimentation on Man* [33], he meant the phrase to be deliberately emotive. For him, "guinea pig" was much nearer to the connotation George Bernard Shaw intended in what the *Oxford English Dictionary* credits as the first metaphorical equation of human subjects with the helpless furry creatures often found in biomedical research laboratories.[3] Pappworth was highly critical of the range of experiments being performed on human beings, the relative paucity of external controls setting limits on doctors, and the lack of informed patient consent. He attributed much of the recent deterioration of doctor-patient relationships, and the disappearance of the 'physician-friend,' to the wedges which human experimentation drove between physicians and their patients.

Yet even Pappworth, for all his passionate concern to expose what he felt was the abuse of human experimental situations, recognized that some such experimentation was so inextricably intertwined with modern medicine that total abolition was neither feasible nor desirable. Human experimentation is and long has been a part of medical endeavor, although, ironically, only relatively recently has it evoked concern comparable to that of animal experimentation, particularly in the English-speaking world.[4] Indeed, some nineteenth-century antivivisectionists ardently believed that human experimentation involved fewer ethical objections since human subjects could understand the purpose of the experiment and might have the option of refusing to participate.[5] Others held that animal experimentation was pernicious, partially because it bred callous unconcern, which made human experimentation even more dangerous.

Despite the recent growth in literature regarding human experimentation, the history of the topic has remained almost unexplored. Most extended works on the subject include a few historical remarks, but those physicians, lawyers, philosophers, and others who are concerned with it have used history only as a backdrop to their more presentist orientation. As a historian I shall excuse myself from commenting on the present. Rather, a brief framework will be presented within which the reader can understand past attitudes and values. While not exhaustive, this framework will indicate some of the variety of cultural, professional, epistemological and technological settings in which human experimentation has arisen.

Five types of Western medicine will be identified. These types are idealizations whose concrete historical embodiments were never pristine. Yet it is possible to see certain historical patterns in them, for they involve differing values which in turn have influenced those crucial determinants in human experimentation: the doctor-patient relationship, the respective rights and obligations of doctors and patients, and the perceived nature of disease and medical knowledge. Ackerknecht has termed four of these types of medicine *bedside*, *library*, *hospital*, and *laboratory* or *experimental*, to which can be added a fifth, *social* ([2], p. 15). These types of medicine will serve as signposts in our survey of two and a half millennia of medical history.

2. MEDICINE AT THE BEDSIDE

Bedside medicine was the method practiced by the Hippocratics, from whom were derived naturalistic explanations of health and disease, and a formal oath which provided the basis for much subsequent Western medicine. The Hippocratic corpus was not the product of a single writer, and the various surviving treatises espouse a variety of explanatory frameworks, of which the humoral had the most lasting impact.

Humoral medicine placed the active agents of health and disease in the liquids or 'humors' of the body, and consequently devalued the solid body parts. Human dissection was not practiced due to religious scruples, veneration of the dead, and fear of corpses.[6] On the other hand, the doctrine of the 'humors' was easily derived from and confirmed by everyday observations at the sickbed: the sweat of fever, the jaundice of liver disease, the haemoptysis of consumption, the quality and color of urine and faeces, and the phlegm of respiratory disorders. Humoral medicine was also holistic, in that the 'humors' could be found throughout the body and were related to more general factors such as temperament, age, diet, and body *habitus*. The Hippocratic emphasis on familiar features of life such as diet, massage, exercise, and bathing encouraged the physician's rôle as advisor and friend, although the stress laid by the Hippocratic Oath on the secret nature of the doctor's knowledge reminds us that doctors have long had a sense of their special social and epistemological identities.

The Hippocratics also developed an explicit notion of the "healing power of nature" (*vis medicatrix naturae*) – the body's inherent tendency to rid itself of excess or peccant "humor" associated with disease [32]. Thus, many phenomena of acute disease (sweating, vomiting, coughing) were viewed as part of the healing process and, as such, were encouraged. "Expectant Therapy" could be in the patient's best interest.

The Hippocratic context consequently was not conducive to an active program of human experimentation. To this general statement, however, three qualifications need to be appended. First, even though Hippocratic medicine was observational rather than experimental (in the modern sense), the word *experiment* is derived from the same Latin root as the word *experience*, and the Hippocratic tradition certainly valued experience. Any encounter between a doctor and a patient can be *experiential* and thus experimental, if only in a passive sense. The Hippocratics, who were interested in the rhythms and natural histories of disease, and in carefully observing sick people, were increasing their own experience. Further, any therapeutic innovation, even if the doctor believed it to be in the patient's best interest, was an experiment.

Second, the medicine of Classical Antiquity was not uniformly practiced among all social groups. P. Laín Entralgo identified three distinct Hippocratic stances: the gentler, highly individualized diagnoses and therapies for the well-to-do, the quick and to-the-point "resolutive" treatment for the poorer freeman, and the tyrannical, mass-market treatment for the slaves. Slaves were often treated by slave-doctors, but the Hippocratics and more élite doctors would also sometimes treat those who could not pay. Indeed, ancient Greek doctors recognized different kinds of obligation to paying and charity patients ([22], p. 45ff). According to the Hippocratic treatise *Praecepta*, their motives were *philanthropia* (love of man) and *philotechnia* (love of the art [of healing]). The latter also embraced the desire to extend knowledge, and easily encompassed experimentation with new procedures and drugs. Historically, the socially and economically disadvantaged have always been principal subjects of human experimentation and an experimentation pattern was already well established in Hippocratic times.

Third, a point closely related to the above is that despite the holistic and observational nature of humoral medicine, certain Greek doctors were much more active in their approach. There are three or four instances in antiquity of human vivisectional experiments. These occurred first in Alexandria under the Ptolemies and involved the famous Alexandrian doctors Herophilus and Erasistratus. According to the Roman writer Celsus, they received their subjects from prison, by order of the king, and vivisected them in order to understand better the "position, color, shape, size, arrangement, hardness, softness, smoothness, relation, processes and depressions" of the internal parts ([26], pp. 160–161). There are one or two other recorded examples of human vivisection in Greek times, and similar situations may have occurred earlier in other Near Eastern cultures such as Persia ([5], p. 6).

In each case the social setting was constant: condemned criminals, who by their crimes had forfeited their rights, were utilized by doctors, who by royal assent were protected from opprobrium. However, human vivisections were isolated, apparently led to no significant findings, and were opposed by many medical writers of antiquity. They are important, however, for they demonstrate that the Greeks faced the sometimes competing claims of individual suffering and medical knowledge and that, in the interest of the latter, even human life might not be sacrosanct. In addition, these medical endeavors remind us that the relatively passive, observational medicine of the Hippocratics did not hold absolute sway in antiquity. For all his admiration of the Hippocratics, Galen was also an energetic animal dissector whose enquiring mind led him to various innovations which, in their early applications, could be deemed experimental. Galen's experimental ideal was to have important implications for future generations.[7]

3. MEDICINE IN THE LIBRARY

Ackerknecht has termed the period from the waning days of the Roman Empire to the early Renaissance "library medicine." Traditionally, it has been described as a retrogressive time, with a high veneration for authority and little evidence of intellectual curiosity. The coming of Christianity fostered other-worldly attitudes, and suggested a spiritual equality between men, as each individual was believed to possess an immortal soul. The church was most significant during this period, for it was against Christian values that a secular medicine was to emerge.

Despite the close connection between the church and medicine, and the theological equality of Christian souls, "rich" and "poor" medicine still existed, for doctors were deemed to have different responsibilities to those of differing stations in life ([22], pp. 56–57). In addition, the receipt of Christian charity often carried with it reciprocal obligations; those who received it were expected to be suitably docile, cooperative, and grateful, thereby demonstrating their worthiness. Philanthropy received its most visible medical testimony in the hospitals, which, though foreshadowed by earlier Graeco-Roman and Islamic institutions, were the most permanent mediæval legacy to medicine. It was in the hospitals that opportunities developed for more extensive observations by doctors, in a setting where those receiving care were objects of charity, and thereby perceived to be under obligation.

Innovation and experimentation were not generally a part of life in mediæval hospitals, whose primary function was to extend service to orphans, the aged, the poor, and the sick, as well as to travellers. Further, the separation of medicine and surgery removed the physicians from arenas involving bloodshed. However, physicians of the period did recognize one particular area in which experimentation of sorts was necessary: the testing of new drugs.

Classical drug properties were described in classical texts, but new drugs – of local plant origin, for instance – had to be independently tested. Arnold of Villanova (d. 1311), for instance, sometimes followed Galen's injunctions by suggesting that a new medicine should be tested first by the doctor to find out its properties. Arnold was always cautious to use himself as the "guinea pig," however by doing so he risked weakening his own body and thereby limiting his public effectiveness. Arnold's contemporary, Bernard de Gordon, suggested a hierarchy of drug testing: from birds, to brute animals, to "those in the hospital," to the "lesser brethren," and then on to others "in order, because if it [the drug] should be poisonous it would kill" ([10], p. 28). But since animals and human beings had different attributes – their body temperatures might have differed, for instance – data derived from animals were of limited value. Indeed, even individual patients differed greatly in temperament and constitution, making hard and fast rules impossible to determine.[8]

The attitudes of Arnold and Bernard remind us of the persistence of social hierarchies and their impact on medical practice, although the research potential of "hospital medicine" only began to be realized from the late eighteenth century. Long before then, however, energetic attitudes quickened the pace of inquiries involving human subjects. Harvey used some human-derived data in his investigations on the circulation of blood. Sanctorius performed a famous series of autoexperiments on insensible perspiration lasting more than 30 years, and published the results in 1614. Blood transfusions were attempted in the mid-seventeenth century, although the results were not successful and the procedure shortly abandoned [39]. Other well-known examples include experiments on the treatment of syphilis from the early sixteenth century; the development of class-specific therapeutics for syphilis: mercury for the poor and guaiacum for the better off; smallpox inoculation experiments on pauper children; John Hunter's inoculation experiment on himself with the pus from a syphilitic chancre; James Lind's evaluation of the efficacy of citrus fruit in the treatment of scurvy by giving different remedies to different groups of afflicted seamen; and Edward Jenner's pioneering investigations of vaccination on an eight-year-old laborer's son, James Phipps. Equally significant, though less frequently cited instances, are: Hunter's animal research on the development of collateral circulation following arterial obstruction before using deliberate arterial ligation as a means of surgically treating popliteal aneurysms; William Cheselden's experiments on cadavers before attempting a new approach to treating bladder stones; and Spallanzani's investigations on digestion through swallowing and retrieving various substances tied to string.[9]

These and similar instances underscore the extent and variety of human experimental settings in the seventeenth and eighteenth centuries. Some involved animals or cadavers; some aimed to extend physiological knowledge, others to increase pathological or therapeutic understanding; and some were performed on volunteers or the experimenter himself, others on patients who might not have been privy to the nature and purpose of the investigation. Though sometimes useful, animal and cadaver experiments were never wholly sufficient in testing new therapies or surgical procedures.

In the absence of formal public guidelines, the relation between the experimenter and his subject varied according to the nature of the experiment and the dictates of the doctor's own conscience. This reminds us that ethical dilemmas are not unique to modern medicine. It is not correct to assume that earlier doctors, with only ineffective or minimally effective therapeutic choices available, faced no "real" issues when deciding between remedies which, by contemporary standards, are worthless. Today, some might judge a doctor who did nothing rather than bleed and purge a feverish patient to be morally superior to an energetic phlebotomist. But historical judgment should not obscure the dynamics of past medical decisionmaking, or encourage us to forget that the Hippocratic injunction, which was "above all, to do the sick no harm," has different practical consequences depending on the historical circumstances. When at the bedside, or when translating the knowledge of ancient doctors into practical patient care, physicians in early modern times observed, acted, and experimented on their fellow human beings.

4. MEDICINE IN THE HOSPITAL

We have already noted that the creation of medical institutions made human experimentation possible on a vaster scale, because large collections of patients extended the doctor's experience, and because these institutions were created for the social classes that had been expected to be "the first on whom the new is tried."[10] "Hospital patients are, for several reasons, the most suitable subjects for an experimental course" ([3], p. 79), wrote John Aiken in 1777 – suitable because of their illness, because of their resulting obligation to society, and because of the power structure within hospitals that created a regulated research environment.

From the eighteenth century onward, hospitals began to acquire a more modern structure. They became known as teaching, research, and service institutions, or as Michel Foucault put it, as establishments transforming benevolence towards the poor into knowledge applicable to the rich. As one late eighteenth-century doctor wrote:

Beneficent gifts will assuage the ills of the poor from which enlightenment will result for the preservation of the rich. Yes, rich benefactors, generous men, this sick man lying in the bed that you have subscribed is now experiencing the disease that will be attacking you ere long; he will be cured or perish; but in either event, his fate may enlighten your physician and save your life ([12], pp. 84–85).

The details of this "hospital medicine," nurtured by the French Revolution and dominant during the first half of the nineteenth century, have been analyzed by Ackerknecht, Foucault, Gelfand, and others. Undoubtedly, medicine acquired greater authority within hospital walls, and the ideals embodied by French hospital medicine had important consequences for our subject. Three aspects of this medicine are worth emphasizing: the conceptualization of disease in terms of lesions, the premium placed on diagnosis, and the therapeutic consequences. While none of these features was entirely new, their impact was synergistic and therefore considerable.

The lesion concept enhanced the pathological significance of solid body parts, particularly tissues as defined by one of the founding fathers of French hospital medicine, Xavier Bichat. Hospital medicine was based on pathological anatomy, and on the way in which lesions defined diseases, thus increasing diagnostic precision. The great monographs of Corvisart, Bayle, Laennec, Louis, and others brought order to pathology, particularly of the thorax and abdomen. These men also revolutionized the practice of physical diagnosis; percussion and auscultation permitted the doctor to gain a much better knowledge of the pathological changes occurring in the living body, thus correlating sign, symptom, and lesion. The "natural history" of a disease became defined primarily in terms of the evolving lesions produced by disease, e.g., phthisis, pneumonia, typhus, and typhoid. Diagnostic acumen became the hallmark of the eminent clinician.

Coincident with these conceptual changes was a devaluation of many traditional remedies. This therapeutic skepticism – extending occasionally to nihilism – was rooted in a critical mentality most famously represented by Pierre Louis' advocacy of his "numerical method": a scale that based medical experience on the number of past cases handled. Therapeutic skepticism was reinforced by the daily hospital routine among the seriously ill and dying, and by the sharp contrast between the new diagnostic and the still traditional therapeutic capabilities. Therapeutic skepticism, so intimately associated with pathological anatomy, contributed to the public image of hospitals as doctors' playthings and further distanced hospital élites from those they treated.

Pathological medicine was the foil to Broussais' physiological approach, and fuelled acrimonious exchanges between Broussais, Louis, Laennec, and the other high priests of physical diagnosis and pathologico-clinical correlation. Today, Broussais is remembered as one of the amusing culs-de-sac of medical history, a monolithic theorist of disease and the man whose influence brought leeching into popularity for two decades. But he also challenged diagnostically-oriented physicians to examine their values and ask themselves if their concern with accurate diagnosis – and roundsmanship – had not eclipsed the more patient-centered therapeutic and caring aspects of medicine. Even if we judge Broussais' therapeutic claims as worthless, it does not neutralize the historical significance of his challenge ([21], [9]).

The ideals of Paris hospital medicine were soon translated into other medical centers: London, Philadelphia, Dublin, and Vienna.[11] During the nineteenth century, hospitals gradually became the temples of clinical research.

5. MEDICINE IN THE COMMUNITY

Medicine has also had another locus besides the bedside, library, and hospital: the community. The growth of the public health movement since the eighteenth century, and the preventive ideals this movement espoused, all contributed to community medicine.

The movement's goals changed over time, but they were generally directed at eradicating environmental causes of disease: over-crowding, impure water, poor housing, insanitary waste disposal, long and arduous work hours, and dangerous workplaces. Social, or community, medicine did not involve human experimentation *per se*, although it did involve comparison of human data with bacteriological and other information relevant to understanding the causes and prevention of disease. In a broader definition of human experimentation, then, even social medicine relied on the investigative use of human subjects. Epidemiological surveys and prospective clinical studies were aimed at elucidating the natural history of diseases or determining the influence of unchecked environmental factors on human health. Screening programs were aimed at uncovering treatable, asymptomatic disease, but they were also primarily investigative, as in determining correlations between age-specific mortality and elevated blood pressure, or dietary habits, and carcinoma of the bowels. Such surveys were and are of great significance to modern medicine; however, they may involve withholding treatment or advice from a patient in the interest of statistically significant data.

In fact, the coming of the statistical world view, championed by Francis Galton, Karl Pearson, and others in the late nineteenth century, had important consequences for the nature of medical knowledge and the extent of human experimentation. Nature speaks to us in statistical language – a language that inevitably reduces the sick person to a case, or the life cycle to a series of probabilities. Statistics may also protect the patient from ill-founded therapeutic claims, and there is an increased likelihood that each individual will at some point be a part of an epidemiological measurement or clinical trial. Statistics may be of use in the hospital or laboratory, but it is in the community – in mass screenings or epidemiologically-grounded advice on eating or smoking habits – that the ordinary person is likely to confront them.[12]

6. MEDICINE IN THE LABORATORY

The boundaries between bedside, laboratory, and hospital medicine were never razor-sharp. Just as the first half of the nineteenth century saw the crystallization of hospital medicine, the second half witnessed the maturation of laboratory medicine. One of its greatest practitioners was also one of its most subtle and eloquent spokesmen: Claude Bernard. Bernard's *An Introduction to the Study of Experimental Medicine* [6] remains a classic statement of the goals, methods, and values of the experimental mind. His comments on human experimentation have often been quoted. Of equal significance is the persuasiveness with which he placed physiology – by which he meant a knowledge of function – at the heart of medical understanding; he did this by simple extensions of knowledge gained from his other two branches of experimental medicine: pathology and pharmacology. These sciences were generally *actively* experimental. The experimenter held constant as many parameters as possible so as to create an environment in which he could understand changes and their causes. Bernard worked primarily with animals, but recognized that medicine was ultimately concerned with human health and disease, and that there were many experimental questions for which human subjects must be the last court of appeal [6]. Bernard's favorite pupil, Paul Bert, conducted a famous series of experiments on the effects of altitude on oxygen pressure and respiration; he began first with dogs and finally extended the testing to include himself and other human subjects. Though of relatively esoteric interest in the 1870s, these experiments subsequently became the basis of aviation medicine [1].

The growth of medicine during the past century has gone hand in glove with the growth of human experimentation, for the actively enquiring mind Bernard sketched has become a common part of the modern medical landscape. Animals, isolated organs, and tissue cultures provide many answers. But diseases without adequate animal analogues, new drugs, and therapeutic procedures, and questions involving higher psychological functions such as pain and subjective responses, are among the areas where innovation and understanding finally must come from man. Legendary events, such as Walter Reed's research on yellow fever, Henry Head's work on cutaneous sensation, or the investigations of Florey, Chain, and other colleagues on penicillin, are simply apotheoses of ordinary facets of modern medicine (see [4], [18], and [27]). Indeed, given the nature of accepted experimental proof and the technological status of contemporary medicine, questions inevitably arise regarding ethical safeguards, the relationship between animal and human experimentation, and the balancing of knowledge against risks. Even arguments like McKeown's that the most significant improvements in health have, and will continue to, come from nutritional, environmental, and behavioral modification, do not seriously challenge the cogency of much basic and clinical research.[13] The present century has witnessed the amalgamation of bedside, hospital, and laboratory medicine into what is known as clinical research.

7. LEWIS AND THE IDEA OF CLINICAL SCIENCE

Spectacular advances in the basic sciences during the second half of the nineteenth century did much to consolidate the relationship between medicine and research in the public mind. Both the physiological concept of disease (that pathology is merely altered physiology) and the germ theory (that many diseases are caused by microorganisms) were easily adaptable to the animal laboratory. Full-time careers in the medical sciences developed before those in clinical medicine, where, particularly in the United States and Britain, clinical teachers were expected to derive most of their incomes from private practice. This meant that clinical research took hours away from time that might have been more lucratively spent.

Clinicians were aware of the relative inequality created by the growth of basic medical science, and of the potential for scientific knowledge to outstrip clinical. The phrase "clinical science" had been used as early as the 1870s by the English surgeon Sir James Paget, but practitioners in medical schools and major hospitals were concerned that the common institutional arrangements hardly encouraged physicians to apply the same vigorous standards of laboratory science to the bedside. Calls for more propitious conditions for clinical research were made with increasing frequency in the years before World War I. In the United States, activity in the area gained momentum with the sweeping changes in medical education associated with the efforts of Abraham Flexner and others.[14] Since its foundation, Johns Hopkins had been at the forefront of newer approaches, and the importance of clinical research was understood at the Rockefeller Institute for Medical Research in New York City, when, in 1910, a hospital was opened in conjunction with the Institute, to allow patients to "be studied with an unprecedented degree of thoroughness."

Wider ramifications of the experimental climate among élite clinicians can be seen in several organizations devoted to furthering clinical research, especially in the American Society for Clinical Investigation, which first met in May, 1909, under the Presidency of Samuel James Meltzer. Meltzer was in charge of the departments of physiology and pharmacology at the Rockefeller Institute. The range of the many achievements in clinical research in America between 1905 and 1945 has recently been chronicled by Dr. McGehee Harvey ([15], [16]). Dr. Harvey points out that many of the early exponents of clinical research in America had received their inspiration from Germany, where full-time academic careers were easier to obtain in the polyclinics, and where deliberate integration of clinic and laboratory had been achieved. In Britain, the situation in the first decade of the century was still the traditional one of part-time hospital consultants pursuing mixed careers of teaching and practice, with clinical research being a luxury sandwiched in by those so inclined.

It was in this milieu that Thomas Lewis emerged. His interest in research dated from his preclinical days at University College, Cardiff. The absence of a clinical school there necessitated his move to University College Hospital (UCH) London, with which he was associated almost continuously from his arrival as a clinical student in 1902, until his death in 1945. At UCH, the surgeon-scientist Sir Victor Horsley particularly influenced him. Horsley was a classic example of the British system, for he had a prodigiously large private practice, heavy hospital duties and an active involvement in public affairs. His research, particularly on the localization of cerebral functions, was usually conducted between five and eight o'clock in the morning [17].

Early in his career Lewis was faced with the problem of earning a living. Despite his prompt advancement up the UCH hierarchy, he also started a Harley Street practice. Clinical research was his real love, however, and his early international reputation as one of the pioneers of modern cardiology put him in an ideal position to receive financial support from the newly-formed Medical Research Council (MRC) and the Rockefeller Foundation. These two sources, one public and the other private, enabled Lewis to devote most of his career to clinical research.

It is interesting to examine Lewis' reflections on the central importance of research on man for modern medicine. Like Claude Bernard, Lewis used his own research as the basis of his philosophy of experimentation. His research ranged widely: from the electrical manifestations of cardiac function, to the physiology of the human skin and superficial blood vessels, to the mechanisms of pain. Lewis expounded his philosophy in a number of addresses and papers, and published it in volume form in *Research in Medicine*, and in a separate book, *Clinical Science, Illustrated by Personal Experiences* ([23], [24]).

Lewis was conscious of his relationship to Bernard, for he felt that "experimental medicine" was the more appropriate phrase to describe his own activities. Bernard's phrase was so associated with animal research, however, that Lewis chose to use the phrase "clinical science" instead. Lewis felt that the success of experimental physiology since Bernard's day had been purchased at the price of loosening its ties with clinical medicine. The result was detrimental to both: "To divide or attempt to divide medical research into ward research and laboratory research is narrow and harmful." Medicine, after all, was about men, and physiological data and disease models from animals, for all their usefulness, were no more than approximations of human systems. In the end, all certain data about the human organism came from man, either through planned experiments, or through the opportunities nature provided in the forms of disease and injury. The methods of clinical science are the same as those of experimental medicine, despite the fact that the ethical dimensions of each are often different. One of Lewis' primary aims was to bring these two disciplines closer together.

Lewis illustrated this mutual dependence of clinical science and experimental medicine through several examples from his own work. For instance, auricular fibrillation had first been observed in patients, and its possible causes examined in the clinical setting by Cushny, Wenckebach, and Mackenzie. It was clinical observations, reinforced by readings in the string galvanometer (the fore-runner of the E.K.G.), which identified the pathophysiological problems which Lewis and others investigated in dogs. The wider problem thus had two aspects: one human and one animal. The animal studies elucidated physiological aspects of the clinical problem; the problem, however, was originally defined by observations made on human subjects. Analogous parallels can be drawn on other aspects of cardiology, pain mechanisms and the nature of the 'triple response of Lewis' (which occurred when skin was subjected to irritation). In these and other major areas of research, human and animal observation and experimentation proceeded together.

Lewis was aware of the ethical dimensions of clinical research, the necessity of informed consent, the competing parameters of potential values and dangers, and the responsibility of the physician for his patient.[15] But he valued the arrangements during his lifetime which created the possibility for more rigorous clinical investigation, for only through clinical science could clinical medicine achieve the certainty which Bernard's triad of experimental medical sciences – physiology, pathology, and pharmacology – had obtained.

Just as the nineteenth century saw the maturation of three kinds of medicine – hospital, social, and laboratory – the present century has seen the coming of age of Lewis' clinical science. There can be little doubt of its importance, but like any human endeavor, it involves human values and judgments.

The Wellcome Institute
London, England

NOTES

[1] Works which I have found useful but have not cited specifically include [19], [14], [21] and [7].

[2] Mellanby's concluding opinion was: "At some stage or another almost every advance in medical science has depended on the use of a 'human guinea pig'." See [31], p. 92.

[3] Pappworth's opening sentence is: "The main purpose of this book is to show that the ethical problems arising from human experimentation have become one of the cardinal issues of our time" ([33], p. 9). The *Oxford English Dictionary* attributes George Bernard Shaw with the first use (in 1913) of "guinea pig" as meaning "a person or thing used like a guinea pig as the subject of an experiment." Shaw's sentence read, "The . . . folly which sees in the child nothing more than the vivisector sees in a guinea pig: something to experiment on with a view to rearranging the world."

[4] For more information on antivivisection see [13]. The case for animal research has recently been eloquently stated in [34].

[5] The issue is discussed in [13], p. 319ff. Of course, most antivivisectionists were also concerned with human experimentation.
[6] [26], p. 162. Some interesting historical parallels between attitudes towards dissection and research on live subjects are drawn in [11].
[7] Galen may be seen as an experimentalist in [30].
[8] [29], pp. 22, 26, 40. I am grateful to Dr. Faye Getz for this and the previous reference.
[9] For a general statement of issues in eighteenth-century medicine, see [8].
[10] The paraphrase is from Alexander Pope's *Essay on Criticism*, line 335.
[11] For more information on hospital medicine in France and elsewhere, see [36], [37], and [38].
[12] See [25] for one aspect of the use of statistics in medicine.
[13] "Although I believe that the solution of most disease problems will come in the future, as in the past, from control of disease origins, I hope I have left no doubt about my recognition of the continued importance of research aimed at increasing our understanding of disease mechanisms" ([28], p. 174).
[14] An interesting gloss on these American developments as they are related to biochemistry in both its laboratory and clinical aspects is found in [20].
[15] Lewis emphasized the tensions between curative medicine and clinical research in this way: ". . . curative medicine deals with one individual, while progressive medicine is collective in its outlook" ([23], p. 3 and [24], p. 18).

BIBLIOGRAPHY

1. Ackerknecht, E. H.: 1944, 'Paul Bert's Triumph', *Bulletin of the History of Medicine*, Suppl. 3, 16–31.
2. Ackerknecht, E. H.: 1967, *Medicine at the Paris Hospital, 1794–1848*, The Johns Hopkins Press, Baltimore, Md.
3. Aikin, J.: 1771, *Thoughts on Hospitals*, Joseph Johnson, London, U.K.
4. Bean, W. B.: 1977, 'Walter Reed and the Ordeal of Human Experiments', *Bulletin of the History of Medicine* **51**, 75–92.
5. Beecher, H. K.: 1958, *Experimentation in Man*, Charles C. Thomas, Springfield, Ill.
6. Bernard, C.: 1957, *An Introduction to the Study of Experimental Medicine*, trans. H. C. Greene, Dover Publications, New York.
7. Brieger, G.: 1978, 'History of Human Experimentation', in Warren, W. T. (ed.), *Encyclopedia of Bioethics*, Vol. 2, Free Press, New York, pp. 683–692.
8. Bynum, W. F.: 1980, 'Health, Disease and Medical Care', in Rousseau, G. S. and Porter, R. (eds.), *The Ferment of Knowledge*. Cambridge University Press, Cambridge, U.K.
9. Canguilhem, G.: 1978, *On the Normal and the Pathological*, trans. C. R. Fawcett, D. Reidel Publishing Company, Dordrecht, Holland and Boston, Mass.
10. Demaitre, L. E.: 1980, *Doctor Bernard de Gordon: Professor and Practitioner*, Pontifical Institute of Mediaeval Studies, Toronto, Canada.
11. Dowling, H. F.: 1967, 'Human Dissection and Experimentation with Drugs', *Journal of the American Medical Association* **202**, 1132–1135.
12. Foucault, M.: 1973, *The Birth of the Clinic*, trans. A. M. Sheridan-Smith, Tavistock Publications, London, U.K.

13. French, R. D.: 1975, *Antivivisection and Medical Science in Victorian Society*, Princeton University Press, Princeton, New Jersey and London, U.K.
14. Freund, P. A. (ed.): 1972, *Experimentation with Human Subjects*, George Allen and Unwin, London, U.K.
15. Harvey, A. M.: 1981, *Science at the Bedside: Clinical Research in America, 1905–1945*, Johns Hopkins University Press, Baltimore, Md. and London, U.K.
16. Harvey, A. M.: 1976, *Adventures in Medical Research, A Century of Discovery at Johns Hopkins*, Johns Hopkins University Press, Baltimore, Maryland, and London, U.K.
17. Himsworth, Sir H.: 1982, 'Thomas Lewis and the Development of Support for Clinical Research', *Pharos* **45**, 15–19.
18. Jefferson, Sir G.: 1955, 'Man as an Experimental Animal', *Lancet* (1), 59–61.
19. Katz, J.: 1972, *Experimentation with Human Beings*, Russell Sage Foundation, New York.
20. Kohler, R. E.: 1982, *From Medical Chemistry to Biochemistry, The Making of a Modern Biomedical Discipline*, Cambridge University Press, Cambridge, Mass. and London, U.K.
21. Ladimer, I. and Newman, R. W.: 1963, *Clinical Investigation in Medicine*, Law-Medicine Research Institute, Boston University Press, Boston, Mass.
22. Laín Entralgo, P.: 1969, *Doctor and Patient*, trans. F. Partridge, Weidenfeld and Nicolson, London, U.K.
23. Lewis, Sir T.: 1934, *Clinical Science, Illustrated by Personal Experience*, Shaw & Sons, London, U.K.
24. Lewis, Sir. T.: 1939, *Research in Medicine and Other Addresses*, H. K. Lewis, London, U.K.
25. Lilienfeld, A. M.: 1982, 'Ceteris Paribus: The Evolution of the Clinical Trial', *Bulletin of the History of Medicine* **56**, 1–18.
26. Longrigg, J.: 1981, 'Superlative Achievement and Comparative Neglect: Alexandrian Medical Science and Modern Historical Research', *History of Science* **19**, 155–200.
27. Macfarlane, R. G.: 1979, *Howard Florey, The Making of a Great Scientist*, Oxford University Press, Oxford, U.K.
28. McKeown, T.: 1979, *The Role of Medicine: Dream, Mirage, or Nemesis?*, Princeton University Press, Princeton, New Jersey.
29. McVaugh, M. R.: 1975, *Arnaldi de Villanova, Opera Medica Omnia II, Aphorismi de Gradibus*, Seminarium Historiae Medicae Granatanensis, Granada and Barcelona, Spain.
30. May, M. T.: 1968, *Galen on the Usefulness of the Parts of the Body*, 2 vols., Cornell University Press, Ithaca, N. Y.
31. Mellanby, K.: 1945, *Human Guinea Pigs*, Victor Gollancz, London, U.K.
32. Neuberger, M.: 1943, *The Doctrine of the Healing Power of Nature*, trans. L. J. Boyd, (n.p.), New York.
33. Pappworth, M. H.: 1968, *Human Guinea Pigs: Experimentation on Man*, Beacon Press, Boston, Mass.
34. Paton, W. D. M.: 1984, *Man and Mouse: Animals in Medical Research*, Oxford University Press, Oxford, U.K. and New York.
35. Pope, A.: 1711, *Essay on Criticism*, in John Butt (ed.), *The Poems of Alexander Pope*, Methuen, London, U.K. (1965).

36. Shryock, R. H.: 1948, *The Development of Modern Medicine*, Victor Gollancz, London, U.K.
37. Vogel, M. J.: 1980, *The Invention of the Modern Hospital, Boston, 1830–1930*, University of Chicago Press, Chicago, Ill. and London, U.K.
38. Waddington, I.: 1973, 'The Role of the Hospital in the Development of Modern Medicine: A Sociological Analysis', *Sociology* **7**, 211–224.
39. Wintrobe, M. M. (ed.): 1980, *Blood, Pure and Eloquent*, McGraw-Hill, New York.

HANS-MARTIN SASS

COMPARATIVE MODELS AND GOALS FOR THE REGULATION OF HUMAN RESEARCH*

Life neither is nor ever has been without risk. But, in the process of history, human beings have managed to reduce a part of their riskful dependency on raw nature by building homes, farms, and machines, by healing diseases and developing drugs to treat them, and by establishing networks for the exchange of goods and information, as well as networks of health care, of regulations, and of social and political contracting for mutual benefit. These developments, however, have not completely freed human life from risk. On the other hand, we have decreased our dependency on nature by increasing our dependency on economic, political, regulatory, cultural, and technological networks. The capacities and potential catastrophes of these new networks have created new forms of risk endangering not only our survival, but also that which each of us, according to his or her preferences, would consider to be a good life. Especially with the most recent developments in high technology and in social, political, and genetic engineering, we have become aware that the quantity of risks has not really changed much when comparing, for example, great natural disasters with great nuclear catastrophes, or subduing people by means of flogging with subduing them by electrical or pharmaceutical measures ([59], pp. 33–4), or life-threatening risks in natural jungles with risks and deprivations in modern jungles of regulations.

Nevertheless, we must understand the history of technology, and the history of political, scientific, and medical institutions, if not as a progressive diminishing of risks, then at least as an increase of possibilities and options, i.e., increasing the recognition and realization of human dignity and liberty. Such a process demands payment for realized – and often unrealized – benefits. Progress will never be free from risks or provide a water-tight solution to all sorts of conflicts. Political philosophers during the Age of Reason were aware of the tension between *libertas* and *securitas* and the difficult task of balancing these two goals of social and political progress wisely.[1] The philosophy of technology, including administrative and medical technology and their assessment, must draw on an awareness of a similar tension between *possibilitas* and *responsibilitas* [101]. The only means available for balancing our enormously increased technical possibilities with our less developed patterns and means of responsibilities are the same ones that have proven more or less effective in balancing liberty and security. Among these means would be a combination of – publicly accepted – social contracting and – publicly, not openly accepted – paternalism.[2] A discussion of the human need for a flexible and wise balance between liberty and security in political affairs, between possibility and responsibility in professional affairs, should also include mention of the importance of a happy balance between *res privata* and *res publica*.[3] While contemporary Western culture seems to lay more emphasis on the individual's rights, and traditional Eastern thought favors the individual's responsibilities towards family and society, modern totalitarian oligarchies seem unconcerned with either one of these two sets of values.

Within the outlined framework of conflicting values in Western democratic societies, *regulation* has become one of the most favored means of diminishing risks and increasing safety and justice. This includes areas of research in which humans are involved as subjects. These conflicting and concurring values of liberty and security, of possibility and responsibility, of controlling on equal grounds and of paternalizing on unequal grounds, and of personal rights and public responsibilities, become apparent in more specific terms when applied to human experimentation. Technical terms used are: research design, informed consent, controlled clinical trials, liability, benefits, selection of subjects, institutional review boards, peer review, therapeutic or non-therapeutic research, accepted standards of medicine, *primum non nocere*, policy guidelines, guidelines for research or regulation of research, compensation, and privacy protection. Concentrating our analysis of comparative methods and goals of regulation in pluralistic societies, we will not discuss human experimentation in totalitarian states since there do not seem to exist any official moral reflections which might lead to regulations other than those in the interest of power-politics (*Machtpolitik*).[4] Nor will we focus on models and goals developed outside of Western culture. Our topic instead will be confined to the following questions: How do open societies, with high standards in scientific capacity and with reasonable or moderate ethical and political standards, appreciate and conduct research in the life sciences? How do they restrict or regulate research in which human beings are involved as subjects? What are the prevailing moral concepts for the existing forms of appreciation, regulation, or prohibition?

These questions will be dealt with in three steps:

First, we will address the means and concepts of *regulation* in public policy and will examine the role of research ethics and medical ethics within it.

Second, we will focus on the means and concepts of *compensation* and beneficence with regard to an individual participating in research as a subject or patient.

Third, we will discuss various aspects of *assessment* of risk-benefit, of accountability, and of regulations and responsibilities.

I. REGULATION

(1) Forms and Philosophies of Regulation

Pluralistic societies have different forms of regulations ranging from constitutions to laws, by-laws, and administrative regulations. There also are codes of self-regulation within the professions or industry, e.g., rules for professional conduct or means of standardization. Violation of regulations may lead to punishment, e.g., a fine, sentence to jail, or even penalty of death being imposed by public authorities. In professional organizations a violation may cause the loss of membership or a fine.

Research or the pursuit of knowledge traditionally was not supposed to underlie regulation. On the contrary, the Age of Reason fought for the individual's right to pursue truth and his or her freedom[5] to expand freely within the borders set by a contract providing for precisely this secured and safeguarded form of freedom.[6] Only the results of applied research and modern technology were subject to regulation whenever safety, liberty, health or life of the citizens were at special risk. The Technischer Überwachungsverein (TÜV) in Germany, the Occupational Safety and Health Agency (OSHA) in the U.S.A., Technical Control Boards or other regulatory agencies in industrialized countries establish general standards, and inspect individual machines, buildings, or procedures to determine whether or not they meet these standards. Examples include structural and fire regulations for buildings, and safety regulations in regard to constructing and maintaining escalators, cars, and aircraft. Road traffic regulations in the United States, the

Highway Code in Britain, and the Strassenverkehrsordnung in the Federal Republic of Germany form one such set of rules. Driver's licenses and regular automobile inspections are related measures to assure safety. Machines or devices assumed to be less hazardous but nevertheless requiring some form of public control need simply be formally registered, not licensed. Such regulations or registrations primarily have nothing to do with progress in technology or science.[7] While only a few authors believe that, in the age of so-called high technology, the process of developing and of applying further progress should be slowed,[8] most positions do acknowledge a public moral responsibility in general, and demand political and moral guidance of science and technology through regulation. Notwithstanding the safety regulations existing in large areas of public life, research in the life sciences until very recently had not been placed under elaborate regulation. Researchers, self-experimenters, test-drivers, test-pilots, and persons on expeditions and discoverers, generally have been expected to assume their share of risk to the point of risking their lives if their actions broaden existing knowledge or uncover new elements, devices, drugs, rare species, gold, or crude oil. They usually have been seen as heroes, even though the risks they have accepted may have been smaller than the risks taken by an average worker in less exciting fields of labor every day. Working in coal mines or in chemical or steel plants poses occupational risks that by and large seem to be acceptable to society; these risks are not supervised by Ethical Committees and are probably seen as compensated by higher wages or other results of bargaining procedures.

One of the first sets of regulations concerning research in the life sciences, i.e., practicing therapeutical and non-therapeutical medical experimentation, as the "Richtlinien für neuartige Heilbehandlungen und für die Vornahme wissenschaftlicher Versuche am Menschen" ([61], pp. 174–5), was issued in 1931[9]. The "Richtlinien" differentiate between New Therapy (Art. 2) and Scientific Experimentation on humans not serving any therapeutic purposes in any specific case (Art. 3), and also require that New Therapy meet "the principles of Medical Ethics and the rules of the medical arts and medical science in its design and realization". The researcher must "analyze and calculate a reasonable (richtige) relationship between risk and benefits", should first conduct tests using animals (Art. 4), and must have the unambiguous consent of the subject or of his legal representative after providing appropriate information. Using New Therapy without consent is only permissible in cases requiring urgent action to save a life or to protect against severe damage to the patient's health (Art. 5). The physician or researcher should be especially careful in calculating the risk of applying New Therapy to children or minors under eighteen (Art. 6), and must accept that "Medical Ethics rejects any exploitation of social distress" (Art. 7). He must be extremely cautious when using living micro-organisms, especially when using pathogenic ones. New Therapy is only permissible when a relatively low risk can be assumed, and when the expected benefits cannot be achieved by other means (Art. 8). In all health care facilities, it is the head physician who bears "full responsibility" – only "with his definite order and full responsibility" should research be performed by another physician (Art. 9). All researchers should prepare reports addressing the goals, design, and realization of New Therapy. They should take into account the question of informed consent (Art. 10), which is to respect the dignity of the patient and follow the dictates of humanity (Art. 11).

Non-therapeutic Experimentation requires an acceptance of the principles applicable to New Therapy, as well as these additional ones: Research is not permissible without consent of the subject, and human experimentation is forbidden unless there was prior experimentation with animals and all possible data from the medical and biological sciences have been assembled. No random research may be conducted. This is especially binding when experiments involve children or minors under eighteen, even if they entail only minimal risk, or in experimenting with dying people. Such research is "unacceptable according to the principles of Medical Ethics and therefore is not permissible" (Art. 12). Finally, the medical profession, and in particular leading clinical physicians, should consider it their professional responsibility and routine medical practice to search for new ways to provide relief and develop improvements in preventive or therapeutic means, in order to replace ineffective current medical science (Art. 13). In the academic teaching of medicine, every possible opportunity must be used to inform students of the special responsibilities connected to New Therapy and Human Experimentation, including also the responsibilities related to publication of research (Art. 14).

The Richtlinien remained binding law through the end of the German Reich in 1945. They were stricter and more detailed than the Nuremberg Code of 1947 and the Helsinki Declaration issued in 1964. The fact that such a governmental regulation actually existed, at a time when the

Nazis were carrying out human experiments in an irresponsible manner in concentration camps, underlines the irrelevance of legal regulations if they are not enforced by the authorities. It was precisely the human experimentation in Nazi camps – judged worldwide to be cruel, unfair, and immoral – which gave states, professional organizations, and other agencies the incentive to issue laws, regulations, or codes of conduct in order to protect "the principles of respect for persons, beneficence, and justice" ([73], p. 4).

The first and most influential of these codes of conduct was a ten-point statement contained in the Nuremberg Military Tribunal's decision in the case of *United States v. Karl Brandt, et al.*, since known as the Nuremberg Code.[10] According to the Nuremberg Code, human experimentation is justified only under two conditions: (a) when the results benefit society, and (b) when the research is carried out in a way that satisfies certain "moral, ethical, and legal concepts". Among the basic requirements to be met, the Nuremberg Code addresses the following five: (1) *Voluntary consent* of the subject involved in the research is mandatory; the "duty and responsibility for ascertaining the quality of the consent" is the personal responsibility of the researcher and cannot be delegated; (2) *Experimental designs* should be laid out carefully, be based on animal experimentation, and should exclude all arrangements, in which there would be an "a priori reason to believe that death or a disabling injury will occur" ("except, perhaps, in those cases in which the experimenting physicians also serve as subjects"); they also must avoid "all unnecessary physical and mental suffering and injury", and should not be "random and unnecessary in nature"; (3) *Highest technical and professional standards* regarding equipment, facilities, skills, and scientifically qualified professionals and staff are required through all stages of experimentation; (4) *Cessation of the experiment* is required, should the subject request this or should the scientist "believe in the exercise of good faith, superior skill and careful judgment" that a continuation of the experiment might result in the subject's injury, disability, or death; (5) *Risk-benefit analysis* should accept no risk that is not exceeded by the "humanitarian importance" of the problem to be solved ([89], pp. 1764–1765).

These five principles subsequently were instrumental in shaping a variety of codes:

(1) *Codes of Medical Ethics in Research on Humans* issued by national

and international medical associations, such as the World Medical Association's Declaration of 1964 and its revised version issued by the Twenty-Ninth World Medical Assembly in Tokyo in 1975,[11] the documents on *Responsibility in Investigations on Human Subjects*, published in 1953 and 1963 by the British Medical Research Council,[12] the *Ethical Guidelines for Clinical Investigation* issued by the American Medical Association in 1966,[13] and the *Richtlinien für Forschungsuntersuchungen am Menschen* of the Schweizerische Akademie der medizinischen Wissenschaften in 1970–71.[14]

(2) *Codes of Research Ethics in the Behavioral Sciences*, such as the *Ethical Standards of Psychologists* of the American Psychological Association (1977), and other national professional organizations such as the Berufsverband Deutscher Psychologen, British Psychological Society (1978), Nederlands Instituut van Psychologen (1976), Berufsverband Österreichischer Psychologen (1976), Polish Psychological Association (1971), Swedish Council of Social Science Research (1975), and the Schweizerische Gesellschaft für Psychologie (1974).[15]

(3) *Internal Regulation Mechanisms in Organizations Funding Human Research*, such as the *Internes Begutachtungsverfahren* of the Deutsche Forschungsgemeinschaft, Sonderforschungsbereich 89 (Cardiology), Federal Republic of Germany, 4 May 1976 ([38], p. 73), or the National Institute of Health in the United States[16] internal *Guidelines* of 1953, 1966, and finally 1967 on *Protection of the Individual as a Research Subject*, the U.S. National Institute of Health *Guidelines for Recombinant DNA Research*, and the British Medical Research Council *Project Grants. Purpose of Scheme and Conditions* (1979).

(4) *Institutional Review*, i.e., Institutional Review Boards (I.R.B.s) and/or Ethical Committees, according to paragraphs 1 and 2 of the Helsinki-Tokyo Declaration,[17] such as those required by the U.S. Senate,[18] and as first introduced by the Royal College of Physicians of London in 1967.[19] Such review boards also are called Ethical Committees, i.e., Ethische Kommissionen, in Germany.[20] These institutionally required boards or committees vary a great deal in the ways they are nominated, the fields they cover, and in the degrees of involvement of lay persons who might be either community representatives, philosophers, theologians or others. The I.R.B.'s "primary concern" lies in the areas of determining "acceptability of the proposal in terms of insti-

tutional commitments and regulations, applicable law, standards of professional conduct and practice, and community attitude" ([89], p. 1778)[21]. Even though various standard models for the constitution of and the topics for these committees have been proposed, it also has been claimed that there would be no justification for "interfering with existing and effective organizational structures merely on the grounds of uniformity" ([108], p. 1010).[22]

(5) *Legal Regulations*, such as the United States Food and Drug Administration's regulations,[23] and the Federal Republic of Germany's *Gesetz zur Neuordnung des Arzneimittelrechts*,[24] the purpose of which is the protection of citizens as consumers, e.g., protecting consumers of drugs by means of providing information and of distributing certain drugs by prescription only, as well as through highest standards of control, including the introduction and control of human experimentation, e.g., in controlled clinical tests,[25] and tests in risky areas of specialized pharmacological research ([88], pp. 265–308), ([103], p. 119).

(6) *Guidelines for International Governmental or Nongovernmental Organizations*, the first of which was the highly influential Helsinki-Tokyo Declaration, and the most recent being the proposed *International Guidelines for Biomedical Research Involving Human Subjects*, jointly endorsed in 1981 by the World Health Organization (WHO) and the Council for International Organizations of Medical Sciences (CIOMS).[26]

In surveying the development of the goals and methods for ensuring the protection of subjects in Human Experimentation during the fifty years from 1931 (Reichsrundschreiben) to 1982 (WHO Guidelines), it is apparent that the goals have remained the same – "respect for persons, beneficence, and justice" – , while the means for obtaining these goals have changed considerably. The Reichsrichtlinien apply the principles of the Hippocratic Oath to the modern world of human experimentation in clinical therapeutic and non-therapeutic research. They (a) emphasize the leading physician's *personal responsibility* for the subject involved and the profession's general responsibility for progress in the healing arts and sciences, (b) establish some *special guidelines* in special areas of human experimentation, and (c) stress strongly the *teaching of research ethics* and ethics of new therapy on all levels of medical education. All later guidelines, heavily influenced by the Nuremberg Code, (a) give *outside review* – peer review, ethical committees or institutional review boards – an important role and often a final say, (b) rank *informed consent* highest among the principles to be obeyed, and (c) *introduce new forms of human experimentation*, e.g., pharmacological test series and controlled clinical trials, in order to improve safety standards.

During the last thirty years the extent of elaborated forms of regulation and involvement by committees or government in shaping and directing research in the life sciences definitely has increased. From this point of view articles entitled "Die Verantwortung muss beim Arzt bleiben" (Responsibility Must Stay with the Physician) [61] or "Sinn und Unsinn von Ethik-Kommissionen" (Sense and Nonsense of Ethical Committees) [96] express discontent with the mainstream of a milieu of regulation.

(2) Regulation and Responsibility

Looking at the three forms available for controlling research procedures – *simple notification*, *guidance* or *recommendation*, and *regulation*[27] – the agent bearing responsibility is different in each case. In the case of simply notifying an agency set up by a professional organization, by a research facility or by a state or federal agency, these would only be informed on the means, data, and goals of research, while leaving total responsibility with the scientist. But registering the design and results of research according to a specific form established and provided by such agencies would make the agencies, not the research person, responsible for the specific selection of information required, and already would lead to a sharing of responsibility for the effectiveness and informative standards of data acquired. Guidelines and recommendations requiring formal procedures, e.g., written acknowledgement of informed consent or specific forms for establishing review boards and the issuance of their reports, and the pre-establishment of a set of material values to be protected, such as informed consent or immediate cessation of the experiment or treatment if requested by the research subject or required by the *primum non nocere* rule, give a higher percentage of responsibility to guiding and recommending agencies. This is especially the case if review committees expressly approve or propose the change of specific procedures. Regulations thus far relieve the individual scientist's or physician's responsibility enormously, transforming moral responsibilities into legal or procedural obligations. On the other hand, elaborated procedures for providing information on research design and lengthy discussion by experienced review boards might increase scientific standards as well as moral awareness, thus indirectly decreasing the risk for subjects involved in research ([52], [29], p. 614f).

The shift of responsibilities and the interrelatedness of moral and legal aspects in protecting human subjects may be demonstrated by arguments introducing and following the *Gesetz zur Neuordnung des Arzneimittelrechts* in the Federal Republic of Germany (1976) [33] replacing formal notification and registration of drugs with material registration and approval. Several prescribed steps within the regulatory process are required to establish "quality, effectiveness, and safety (harmlessness)".[28] The idea of licensing drugs according to established and controlled evidence of "quality, effectiveness, and safety" is widely accepted as morally justified, even required by public social and political responsibilities. Controversial, however, are the specific means by which the general goals of a "fair and wise risk-benefit relationship" and "high standards of medical science" are to be achieved. Arguments center around three issues:

(a) the "actual state of the medical science" (Stand der wissenschaftlichen Erkenntnis),
(b) "controlled clinical trials" (klinische Prüfung), and
(c) "liability" (Haftung für Arzneimittelschaden).

(a) The *actual state of medical science* (Stand der Wissenschaftlichen Erkenntnis) and its true knowledge, according to the Arzneimittelgesetz (AMG), must be determined by a committee appointed by the government. Knowledge of the state of the art within this committee serves as a measure for setting up, controlling, and evaluating the results of controlled clinical trials and for evaluating additional information provided by research-based pharmaceutical companies. It has been mentioned that the "present level" or "actual state" of medical science, as that of any other science in any period, has to be viewed as a "state of controversy", and not a "state of knowledge" (Stand des Wissens) ([64], p. 355f)[29]. Kriele calls the tendency expressed in the AMG a Scientification (Verwissenschaftlichung) of the art of healing, threatening to punish those whose professional performance does not fit into the governmentally determined scientific standard, and who use outside methods or not widely or uniformly accepted medical arts.

How could "the present level of medical science" be defined in court? The Verwaltungsgericht Minden[30] in 1975 dealt with the case of a cancer patient whose medical insurance refused to pay for the cost of a specific drug which, according to the insurance agency, did not meet the current standard of medical science. Two experts, both clinical practitioners, testified that they had had positive experiences in using the drug, while a third expert, director and senior scientist at the Deutsches Krebsforschungszentrum (national cancer research center), declared that the effectiveness of the drug could not be established. He based his view on the fact that this specific drug never had been tested in any of the major cancer research centers of the world because the "intellectual-philosophical background", the anthroposophical Weltanschauung that led to the development of the drug, "is scientifically and medically groundless". With the experts voicing two opposing opinions, the court had to choose between clinical experience and accepted official medical research. In another case, Oberlandsgericht Hamm[31] rendered an opinion stating that "only scientific methods taught by a number of well-known university professors should be accepted". This opinion thereby predetermined that only a special group within the medical profession, namely teachers at medical schools, would have the expertise to define "accepted standards" of medical science.

Under the influence of Kriele's critical position, the Sub-committee of the Deutscher Bundestag (German federal parliament) declared that the "present level of medical science" as a legal term should be defined as a widely accepted "core of methods" (Kernbereich). It also strongly recommended that the government should not favor or set back specific drugs or methods by means of defining "scientific standards" ([50], p. 1562). On the other hand, the *Arzneimittelgesetz* (AMG, par. 55) requires the establishment of an Arzneibuch (pharmacopoeia), "collecting accepted pharmacological knowledge in regard to quality, control, storage, distribution, and labeling" of drugs, and its change or supplementation by means of regulations (Rechtsverordnung) if necessary to ensure an orderly supply of drugs. Kriele, in quoting the European Community's Guidelines,[32] which call for non-registration of drugs only in the case of non-efficacy, demands a "redistribution of the burden of proof" (Beweislastverteilung) ([64], p. 355f). There should be no further regulation if efficacy and/or harmlessness are demonstrated. Otherwise, according to Kriele, the government would not respect the constitutionally guaranteed freedom of research and science,[33] and would make science subject to bureaucratic regulation. He furthermore asks for the replacement of the regulatory system with a system of punishing the abuses of freedom and professional expertise.[34]

Jaffe [56] takes a similar view in stating that existing common law would be able to handle all possible cases dealing with human research. For instance, physically touching a person without his or her consent is an "assault". The manipulation of a person's conduct by misrepresentation or breach of a fiduciary relationship is "fraud", and careless conduct of an experiment is "negligence" against which the subject may take action if injured. Thus, common law and its expanded laws can deal effectively with all legal conflicts that might arise between scientist, subject, and society. Also, none of the regulations criticized by Jaffe and Kriele can fully cover the many borderline cases between human experimentation and accepted or routine medical practice. Every routine practice is partially experimental in character. Diagnosing a patient's disease or needs, anticipating the result of a specific treatment according to previous experience or labeling, making judgments on side-effects, as well as other considerations related to the very individual situation of each patient, are, in fact, opinions based always on a trial and error system. This is especially true if therapy is innovative – and any therapy should be sensitive, aware, and innovative according to the patient's needs.

In order to deal more effectively with borderline cases, Robertson ([92], p. 16–45) divides innovative therapy into two groups. Within the first group, physicians seek primarily to serve the patient and only indirectly gain research-related information. In the second group, research is the primary pursuit and only indirectly, if at all, are the individual patient's needs considered. However, such a differentiation is convincing only in theory. Robertson recommends that the medical profession be encouraged informally to "develop clearer standards" in using innovative therapy, and that, instead of controlling mechanisms, there should be a shifting of the "burden of proof to the defendant physician" if he or she is accused of malpractice or violation of professional standards. In favor of the AMG, it has to be mentioned that drugs prepared on the basis of not widely accepted medical theories, but shown to be harmless through experience, do not have to pass the entire control and approval procedures to gain a license. They simply have to be registered by notification.[35] It also is concurrent with the scientification goals of the AMG that pharmacological and clinical control boards are not ethical committees in the strict sense, since they do not include ethicists, philosophers or theologians.

Given the difficulties in defining true knowledge and the actual state of medical science, including their legal consequences, there is no position that is absolutely persuasive in providing a watertight definition of the boundary between human experimentation and routine practice in medicine. This is especially the case for the psychiatric and other behavioral arts and sciences. Instead of contrasting human experimentation with routine practice, it might be better to differentiate among three forms of scientific certainty and consequently three forms of predictable efficacy and safety in treatment:

(1) In cases in which the physician or scientist has established experience, in which the treatment or drugs used have a proven record of predictable results, and in which the analysis and diagnosis of subjects or patients have left only minor questions on how they would respond to treatment, a low degree of experimental uncertainty and a highly predictable degree of efficacy will be found. Many, though definitely not all, so-called routine cases would fall into this category.

(2) A higher degree of experimental uncertainty and a lower, but reasonable, degree of predictability of results will occur in cases in which the record of proven evidence of the scientist's or physician's experience and of the drug's or treatment's efficacy is not yet established, in which the diagnosis is not clear enough, or the subject's/patient's reactions are not highly predictable. This category would encompass a larger number of cases of so-called routine practice, and would also include innovative therapy, as well as many cases of new therapy and low-risk human experimentation. By and large, the majority of cases involving therapeutic treatment and experimentation in the life sciences would fall into this category.

(3) Finally, there are those crucial cases in which the actual state of true knowledge and established experience, the means for reasonable analysis or diagnosis, and the degrees of predictability of the subject's or the environment's reactions are so low and poor that therapeutic treatment as well as non-therapeutical research would

carry high risks. Among cases in this category would be the application of drugs or devices which have never or not sufficiently been tested, and experimentation in areas in which high risk to the subject involved or even to life in general could not be excluded, e.g., in nuclear experimentation, in genetic engineering, or in bacteriological research.

Differentiating among these three degrees of high certainty, reasonable certainty, and uncertainty in life sciences – in conjunction with low and acceptable risk, medium but reasonably and responsibly justifiable risks, and high risks with individually probably not responsibly acceptable consequences – would make it unnecessary to define true knowledge versus experimentation in legal terms. Rather, this incorrectly formulated problem would be replaced with arguments on the proper forms of risk assessment and risk acceptance. The highest-risk category, according to our Western understanding of security and liberty in society, would be subject to legislative responsibility and measures, and consequently in part be subject to governmental regulation or supervision. Risks in the two other categories, however, would be best controlled by professional responsibility, which might include some kind of "intra-professional regulation" ([112], p. 88) but would not require the absurd undertaking of defining by majority vote the "actual state of medical science".

(b) *Controlled Clinical Tests*. Controlled Clinical Tests (C.C.T.s) represent human experimentation that is introduced and required by law and does not necessarily benefit the subjects involved. They are required by law in most Western countries prior to filing a New Drug Application. It has been estimated that only one out of ten of all drugs tested in this area of human experimentation – in the U.S. under the Investigational Drug Exemption, in the F.R.G. according to par. 22.2.3 of the AMG – will finally be chosen for marketing. It also has been estimated that roughly 300,000 persons are involved yearly in experiments with ultimately not approved drugs ([19], p. 234). Lengthy and costly C.C.T.s, required by law, lead major drug companies to concentrate on a more limited number of drugs with an anticipated large market while neglecting research in markets anticipated to be smaller. This also leads companies to have essential parts of drug or device testing conducted outside the United States ([19], p. 232) or other

Western countries. Already fifteen years ago, Cavers called the U.S. Food and Drug Administration's approval procedures wasteful because of the needlessly large volume of clinical tests. According to Cavers, these procedures also discourage research because of the costs involved, and change "the investigator's obligation to his patient". Consequently, "the process of research and innovation in drug therapy" is impaired, causing "a major shift in the responsibility for carrying out drug investigations" ([19], p. 242) from open and free pharmaceutical research and clinical investigations to governmental regulation of research and science, and from professional responsibility to legal and regulatory obligation.

In contrast, the basic idea of C.C.T.s according to the World Health Organization is "a carefully, and ethically designed experiment" designed to show "whether a drug has a useful effect in treating or preventing disease, and to evaluate it, in terms of efficacy and toxicity, in relation to other therapy" ([124]). Since efficacy in therapy is both a goal in medical science and in medical ethics, the philosophy of CCT generally desires to lower the risks of new drugs intended for general use by discovering possible risks and benefits in human experimentation involving a limited number of persons. New moral, as well as professional, issues arise owing to the design of the C.C.T. when groups of patients are treated differently; this is especially crucial when using placebos. While the U.S. Food and Drug Administration requires C.C.T.s, the Federal Republic of Germany's AMG also has simply made C.C.T.s the favored form of providing evidence of efficacy. The C.C.T.s have been strongly supported by the WHO and by governments since they permit the maintenance of high safety standards for the general public, and they are generally accepted by pharmaceutical researchers because of the highly developed standards in statistics and other methods of sophisticated research design.[36] But the use of C.C.T.s also has been severely criticized as ethically irresponsible in regard to the treatment of the human subjects involved, and in regard to an unfair and illegal, perhaps even unconstitutional,[37] allocation of risks to one or more groups of individuals with anticipated or assumed benefits for others not participating in the research. Also, the relevance of the statistics standards used in the C.C.T.s has been questioned (see, for example, [14], p. 45).

Martin Fincke argues that (1) possibly fatal results which would not occur otherwise are accepted risks in the C.C.T., (2) the ethics of *primum non nocere* are violated, (3) it is morally impermissible to give responsibility toward society in general a higher ranking than responsibility toward the individual subject in dangerous clinical research, (4) should the rate of fatalities in the group treated with test drug be higher than the rate of the other group, the physician in charge should be sued for killing, and (5) governments and drug administrations favoring C.C.T. are aiding and abetting criminal acts of killing or injuring individuals. Fincke also states that careful and responsible innovative therapy used in introducing and testing new drugs would be the most ethically and professionally responsible method.[38] Such ethical arguments against C.C.T., however, have been superseded by legal considerations. The European Community's Recommendations of 1975 make it obligatory to transform the request for C.C.T.s into national law.[39] On the other hand, it is worth noting that the WHO's recommendations still keeps the moral question open, by calling C.C.T.s necessary, while at the same time stating that it should always be recognized as being unethical to introduce drugs not "adequately" tested: "The ethical problem is not solely one of human experimentation; it is also one of refraining from human experimentation" ([51], p. 181f). Ulrich Abshagen complains that "legal formalism is not a sufficient criterion for having solved the crucial problems of medical ethics. This includes the fact that the physician may not circumvent his essential responsibility for the sake of the patient by delegating such a responsibility to the patient in requesting his written assent" ([1], p. 10).

In 1977, a European workshop was held in Sesti Levante, Italy, in order to survey the costs and benefits of governmental regulatory agencies, especially with regard to C.C.T.s. The participants, among them physicians and pharmacologists of highest international repute, generally understood the existence of regulations and regulatory agencies as "an important element in the care of public health" and "in the best interest of the individual, of the community as a whole, and of the providers of new medicines". But the hope for a "guarantee of safety", and consequently a steady tightening of the regulatory screws was criticized as being the wrong approach ([26], pp. 233–34).[41] Specifically, the following policies were accused of causing the dwindling of pharmaceutical progress during the last ten years: (a) unrealistic standards of safety, (b) demands for excessive human experimentation resulting not in an increase of safety, but instead in delays, (c) repetition of preclinical and clinical testing, even if already conducted in other countries, (d) thereby channelling research into a limited number of fields, where high costs could be recovered more easily, and (e) neglecting research on less common diseases. In its public statement, the workshop's participants, recalling that "medicines can never be entirely safe", expressed their concern about present methods of testing for safety as consuming "time, money, and manpower without a corresponding increase in safety". The statement also proclaimed a need for "revising our methods of assessment of medicines", especially through (a) "more rational but less extensive laboratory studies, without unnecessary multiplication of detailed clinical trials before registration", and (b) "by closer and more extensive surveillance of medicines after they are available for general prescription". Striving for "absolute safety", which is unattainable, does "more harm than good" ([26], p. 237).

(c) *Liability* is related to responsibility. Shifting personal moral responsibility of the practitioner or scientist by means of law or other regulation to committee bodies or regulatory agencies also means a shift in legal liability, even though such liability is not always accepted by these agencies or cannot be allocated effectively. It has already been mentioned that the traditional British-American common law would be able to handle new cases in human experimentation, and that it would be well prepared to develop case law in these highly technical fields of the life sciences.[42] The same would be true for law based on a system of principles rather than on case history, such as German law. The terms "Körperverletzung" (bodily injury), "Einwilligung" (consent), and "Unterlassung" (omission) defined in the Strafgesetzbuch (penal code) are sufficient for dealing with all possible cases in physician-subject/patient relations [52].[43] The AMG has made the allocation of liability quite complicated, if not impossible. Generally, legal action cannot be taken with respect to existing legislation unless it is in violation of the constitution. Nor can the administrative agencies responsible for implementing the regulations and provided with discretionary power be prosecuted, since they act only indirectly by relying on the results of committee work. Committees only consult and propose; they are not legally responsible for legal registration or regulation. Also, it is nearly impossible to sue individuals on a personal basis, even when they have been heavily involved in shaping committee decisions. Only the agent appointing a committee, e.g., a regulatory agency or a hospital, can sometimes be sued for damages. Very seldom and only in extreme cases of disloyalty or negligence does the employer have the right for recourse ([30], pp. 616–61) to individual committee members. Given the factually hopeless legal situation in regard to liability, the AMG requires that in human experimentation in C.C.T.s, in addition to (a) the written consent of the subject, there be (b) acceptance of responsibility by a physician with at least two years experience, (c) pharmacological-toxicological controls prior to C.C.T., (d) detailed information to the responsible physician on results of the pharmacological-toxicological test given by a scientist involved in the testing, and (e) an insurance of at least 500,000 Deutsche Mark (ca. 300,000 U.S. dollars). Paragraph 40.3 of the AMG states that there be no further action for damages if the insurance company pays.[44] The situation becomes even more confusing when the AMG requires that the drug producer be held liable, whenever labeling or other forms of information do not meet the standards of medical science or when damages exceed norms that are acceptable according to medical knowledge and are caused either by development or production of the drug.[45] If the person who suffers damages is jointly guilty, e.g., in not having followed instructions given, he or she also is partly or fully liable for damages incurred. This all serves to demonstrate the extent to which personal liability together with personal ethical responsibility has been replaced by a form of anonymous committee or administrative liability and by legal formalism. It is self-evident that such a "legal hegemony" not only undercuts personal responsibility; it also might effectively create a new form of "complicity in allowing patterns of professional dominance" ([56], p. 198).[46]

(3) Overregulation and the Wisdom of Regulation in Public Policy

The Arzneimittelgesetz (AMG) is only one example among other cases underlining similar issues in ethics and in public policy that have led to the philosophy of specifically introducing and regulating human experimentation in the life sciences. The philosophy of regulation in other areas of the behavioral or medical sciences and the means and goals applied in regulation procedures in various other Western countries create similar ethical problems and administrative liabilities. These philosophies and methods of regulation are not competitive but recurring, not really different but similar (cf., [66]). There naturally do remain differences related to specific national traditions or to a different length or intensity in the national process of increasing regulation;

however, "international standards" have already been proposed ([24], pp. 68–71).

Originally a very special moral problem, the use of prisoners of war in research, in which they may be humiliated, injured, or killed, has developed into a general field of moral concern leading to legal and administrative regulations for a small and very specific area of modern society. Traditional codes of professional conduct, such as the oath of the Hippocratic physician or of the Christian nun-nurse, emphasizing personal responsibility curiously enough did not directly influence the new national and international Codes and Regulations. Within slightly more than thirty years, medical ethics and research ethics have been solidly embedded in administrative regulations and have acknowledged the need for approval by committee, if not the total replacement of personal responsibility by regulatory obligation. The ethical issue has *shifted* from ethics in research and medicine to ethics in legislation and administration. By and large, problems of *professional ethics* have become problems of *public policy*. Questioning such a development does not automatically mean to question the general interest and right of the public to regulate ethical conduct "as a part of its social contract with the profession" ([116], p. 90), but it does question the prudence of balancing public responsibility and personal responsibility and of replacing primary personal responsibility and liability with lengthy regulatory procedures and forms of shared or limited liability. As has been mentioned earlier, other areas of occupational life with risks comparable to or even higher than those in medical research and new therapy, such as certain working conditions in mining or chemistry, and even more in test flying, mountain climbing, skiing, boxing, or acrobatics, are much less regulated than human experimentation in the life sciences.

The goals of such regulations in the life sciences encompass a fair distribution of risks and benefits (*distributive justice*), the general decrease of risks (*safety*), and the protection and encouragement of progress in the medical and behavioral sciences and arts which only can be achieved in a climate of *liberty* and innovation. We therefore must ask whether these three goals can be adequately achieved in a regulatory milieu. Moral aspects, which have not yet been resolved in a fair and appropriate manner within this milieu of regulation, seem to include (a) a confusion with regard to responsibility and liability in many of the recent laws and regulations, accompanied by an unawareness of the ethical and legal questions of compensation, (b) an overburdened principle of informed consent which in many cases may lead to formalisms of shared responsibility between physician and patient or scientist and subject, and finally (c) an unbalanced distribution between risks and benefits, especially in randomized clinical tests. All three aspects address professional ethics as well as the ethics of administration and legislation. There seems to be a tension between the principles of *responsibility* and *regulation* which needs further discussion and a fairer solution, probably in some form of deregulation and a new appreciation of direct and personal responsibility. Such a development might already be on its way. Its first indication in the U.S.A. was the January 1981 *Final Regulations Amending Basic H.H.S. Policy for the Protection of Human Research Subjects*,[47] proposing substantial deregulation of FDA drug applications, cutting an estimated seventy percent of paperwork, and easing the number of tests on humans; also, the AMG is under legislative review since 1987.

II. COMPENSATION

The important requirements of informed consent and respect for human dignity outlined in the Nuremberg Code and described by Hans Jonas for creating morally responsible forms of human experimentation ([58], pp. 1–31) did not necessarily have to lead to a system of protecting the individual's health and dignity from unwarranted damage by means of increasing regulation. Given a favorable cultural environment, they could have led to a new appreciation of personal responsibility and compensation. In an open market of contracting by voluntarily acting persons there would be many forms of compensation. For the prisoner, assuming he is still given the full or partial legal capacity of contracting, forms of compensation might include (a) an active means of compensating (if not repaying) society for damages caused, which might even lead to a rearrangement with regard to his sentence, or (b) more rewarding conditions in intellectual and social terms as well as amenities during his imprisonment, including better health care. For the individual able to identify himself or herself with the service possibly rendered to suffering present or possible persons as a result of a risky self-experiment or experiment, such an identification would in itself, as Jonas emphasizes, be a great compensation – with or without prestigious medals, awards, or titles bestowed by important committees or organizations or by highest national or international political authorities. The most common forms of compensation, however, are those of the job market, which are assured in any specialized work, such as heavy physical or difficult intellectual work, highly responsible labor, and risky occupational performance in general. For non-therapeutic, and for a part of therapeutic research, there would be clear-cut job descriptions available that could encourage, for whatever reasons, an increasing number of people to practice such an *occupation* in their professional work. Required perhaps to pass aptitude or physical tests from time to time, persons involved professionally on a full time or part-time basis in human experimentation would gain valuable professional experience. Such a professional group of test persons might seek unionization; they also might lobby for safety regulations or higher forms of compensation or risk insurance, as well as press for institutional review boards focusing on ethical, political, medical, or economic issues, although space pilots, uranium or coal miners, stunt persons, and acrobats do not seem eager to have ethical instead of bargaining committees. In addition to the dignifying contribution to the healing sciences in therapeutic research involving patients who suffer or will suffer in the future, other direct forms of personal compensation, such as (a) a change in treatment, however risky, (b) a closer relationship with the treating physicians and the staff in large-scale hospitals, (c) a higher standard of professional treatment, and (d) the medical staff's greater awareness of possible risks always present in medical treatment.

As a result of the regulation process all parties seem reluctant to have prisoners involved as subjects in the research being conducted [74].[48] Also, if *prisoners* have lost their citizens' rights as part of their sentence, they would not be able to contract freely under most legal systems. Somewhat self-righteously, the American Medical Association in 1952 informed all federal and state agencies that "persons convicted of various crimes should not qualify for pardon or early parole" and should "be considered ineligible for meritorious or commendatory citation" ([65], p. 264f). A more serious argument in favor of excluding prisoners from participation in experiments is that they might be influenced by promises or hopes of reward, so that they would not be in a position of free consent ([65]; [41], pp. 687–92). Lasagna provides a major argument in favor of prisoner involvement. He states that "participation" in a variety of forms would enrich the prisoner's life, might result in resocialization, or put him or her in a position of contributing something worthwhile to society ([71], p. 265).[49] In contrast, the U.S. National Commission for the Protection of Human Subjects of Biomedical and

Behavioral Research only approves regulated and supervised behavioral and sociological research related to prisons or institutional structures and to prisoners' health conditions. With the exception of research which satisfies the conditions of equity and guarantees a "high degree of voluntariness," other forms of research involving prisoners generally should not be conducted or supported. The separation of research participation and parole considerations must be assured ([54], [79]). A paper prepared for the National Commission by Roy Branson argues that the principle of justice would require that prisoners be paid the same as other volunteers, while the principle of uncoerced consent would require that they not be paid more than for other prison jobs ([11], p. 51f). Marx Wartofsky, in comparing research subjects' compensation to other forms of high-risk wages, states that in mining, construction, or chemical manufacturing, health is "sold" only as a "by-product" of other services, but that in research it is the "primary commodity" sold, just as a prostitute sells his or her body. His general recommendation is to "minimalize the exploitive elements which 'commodify' the situation" either by stopping the payment of prisoners and conducting only research which could also be conducted with unpaid volunteers or by following Hans Jonas' recommendation of having the most valuable (rather than the most exploitable) members of society undertake the risks involved in human research. Wartofsky also recommends that all research subjects "be organized, educated as to their rights, and represented at all levels of review", measures which again would serve to de-commodify the relationship between researcher and subject ([79], pp. 53–5). The reluctance to involve prisoners in research is influenced heavily by the Nuremberg Code tradition, but it also represents a general, not fully rationalized understanding that punishment by imprisonment is the denial of participation in society, including the denial of participation in the market of goods and rewards.

In looking at one other special group of persons needed in research – children – the issues become even more complicated. There is a general understanding in our societies that *children* should be given special protection – for example, by placing legal restrictions on liquor consumption, sexual activity, education, driver's licenses, and on the capacity for contracting. Such a concept seeks to provide complete protection of the health, well-being, and personal development of children. As a result, their use in human experimentation is frowned upon. On the other hand, there are certain pharmaceutical or clinical tests concerning pediatric medicine that cannot be carried out on adults. Facing this dilemma, all civilized countries have regulations providing for (a) minimalization of the use of children as subjects, (b) protection and control of protection, at least by means of written parental consent and review boards, and (c) nevertheless, the opportunity for biomedical and behavioral research on children for their individual benefit, and for the benefit of pediatrics generally. Society as a whole also may benefit from research on children, e.g., in the area of vaccination research.[50] Special ethical problems have been addressed with regard to the use of fetuses or of children who have suffered brain death. The controversial nature of these issues is related to the diversity of philosophies concerning the state of a fetus or of a human being after brain death. Fundamental positions are quite diverse, while legal instruments and administrative experience are limited. Further discussion is needed which might either ease or complicate the moral and legal issues involved.[51]

The contrast between the legal and moral situation of prisoners, who – at least theoretically – would have the capacity to contract, and the general incompetence of children to contract demonstrates the limited use of the open and fair market of contracting and compensating. The same natural limitations do exist, however, in other areas of life where persons, who are incompetent for one reason or the other, and children are involved. Not natural, but legal or social limitations prohibit prisoners and others from taking full advantage or participation in many areas of life, including human experimentation. Given the special moral and legal aspects of groups limited in their capacity to contract freely, the concept of compensation continues to be rather controversial. Even though both case and criminal law systems provide a comparatively easy means for taking actions against malpractice, negligence, deception, or lies, the informed consent maxim may cause a legal situation in which no liability claims can be pressed by the subject, who finds himself or herself injured or damaged and claims compensation against the researcher and his or her sponsors. As Paul A. Freund puts it, informed voluntary consent is "the legal requirement for legitimate liability-free experimentation" ([43], p. 321).[52] This makes the informed consent principle useless or of limited relevance in cases in which prisoners, children, the mentally disabled, or the socially needy are involved; this is especially true in cases of no-fault damages. The current legal situation in the United States provides for no general compensatory justice in cases of no-fault damage. Since research subjects are not employed in a normal sense, there is no workmen's compensation or, if the research is sponsored by the government, no Federal Employees' Compensation. The German AMG requires, as mentioned earlier, a small insurance coverage for legally required human experimentation in C.C.T.s, as do the Richtlinien of the Schweizerische Akademie der medizinischen Wissenschaften ([104], p. 617). In double-blind tests there is *ex definitio* no personal penal responsibility between the individual subject and the researcher. If liability may be based only on fault, then the only way to obtain compensation would be to prove that the research design is faulty.[53] Such a research design, however, may have been approved or changed by an Institutional Review Board so that the Board's share of liability would also have to be determined. It is surprising, though perhaps explainable by the predominant focus on the informed consent maxim, that the compensation aspect is totally underdeveloped in human experimentation.

The most common and most beneficial way to ease the risks involved in many areas of modern life and to obtain a fair or reasonable compensation for losses or damages is by means of the *commercial insurance business*. Insurances can cover damage or loss of tangibles, life, or health, ranging from ships and their freight to the legs of a female movie star and vacations ruined by rainy weather, totally or up to a certain amount at risk. The insurance business has a long tradition and high standards of expertise in risk assessment, precision in coverage, cost competitiveness, and promptness in providing compensation. The professional experience of underwriters in even the most arcane areas of risk analysis and procedure design for risk reduction would make these companies a highly desirable partner in venture assessment and planning, as well as in analyzing causes of loss or damage incurred, beyond the role of mere contractors of insurance. Insurance services should be warmly welcomed in the risky and complex area of human experimentation and its jeopardized forms of personal responsibility. Freund stresses that the actual state of moral reasoning with regard to human experimentation – and generally in social philosophy – is characterized by two groups, one emphasizing "responsibility, blameworthiness, rewards, and penalties for behavior", the other "stressing security for the victims against the impersonal doom of modern life" ([43], p. 322). Clearly, the most practical and moral economical solution for both parties would be to assume an insurance policy. Ever since Lloyd's of London began accepting insurance coverage for values lost or unrealizable gains, the probabilities for fair and reasonable success have increased, while blaming the Gods for misguidance has decreased. But the moral question of whether or not in a certain business or transaction the individual has acted fairly or morally correctly – the question of fault – will not simply disappear and cannot be compensated for just with money. It has been suggested that "a compensation fund supported out of the federal tax structure" should be established to provide redress in case of no-fault damage ([2], p. 648). However, such a proposal would only be adequate in a highly regulated experimental environment in which the regulator, i.e., the government, would be the agent reliable in paying for damages and providing compensation. The *International Guidelines* issued by the WHO and CIOMS state that "natural justice demands . . . automatic entitlement to reasonable and expeditious compensation for any injury sustained as a result of participation"; the Guidelines insist on full compensation, in case of death to be given to dependents, and do not accept any waiver of the right to compensation ([126], pp. 19 a. 32f).

Since insurable damages are quite rare, ranging from less than 1 percent injured in non-therapeutic to 10 percent in therapeutic research, they represent only one aspect of moral, legal, and risk-benefit issues in human experimentation.[54] The regulatory approach so far has been to a large extent inflexible, even unaware, of the need for compensation of personal injuries. The compensation approach, in contrast to the regulatory approach, seems to be in a better position for handling the injury issues, including those of no-fault, and for providing a more favorable climate for freely contracting research subject positions. In addition, it is more flexible, as well as consent-oriented in allocating risks and benefits, and probably more effective in pursuing the valuable goals of human experimentation. On the other hand, compensation and insurance, in themselves means of compensative justice, are not ends in themselves and should not replace moral debates with bargaining tables.

III. ASSESSMENT

The guidelines for the protection of human subjects in biomedical and behavioral research and the various forms of regulation intend to prohibit previous or probable misuse of subjects involved. They, in particular, want to protect those belonging to special social groups, such as prisoners and prisoners of war, or poor, uneducated, incompetent, and hospitalized persons. Even though cost-benefit calculations and the assessment of means and goals have been favored recently and have been methodologically refined in utilitarianism, any major philosophical or religious position needs and, indeed, uses assessment devices in flexibly applying theory to practice, with the exception of fundamentalist dogmatic views.[55] Existing forms of allocating risk and benefit in human experimentation by means of regulation are a result of assessing and changing previous distributions of risk and benefit understood to be not permissible or justifiable. Now, after nearly two decades of experience with various and growing forms of regulation, the *regulatory maxim* itself is due for reassessment and re-evaluation. A few preliminary thoughts may show in which direction such an assessment might lead: (1) the revision of the no-risk dream in biomedical and behavioral sciences, (2) the reassessment of means actually used to minimalize risks generally, such as regulatory and reviewing boards, and the informed consent principle, and (3) a reallocation of risks with regard to expected benefits.

(1) Since life in general is not risk-free – and if it were, there would be no place for adventure, progress, success, satisfaction, responsibility – how can the medical sciences be less risky than other areas of life?[56] Given this situation, it would be foolish to require higher safety standards for human experimentation than is presently the case in accepted or routine medical, psychological, social, or educational treatment, political decision making, or in other areas of professional life. This insight should support a philosophy of minimizing regulation and of favoring responsibility, compensation, and insurance within a wider framework of moral and technical risk management.

(2) Existing risk-minimizing devices are regulation and review. The development of regulatory procedures and review techniques in public policy, as well as in professional organizations and research facilities, has been more or less simultaneous. The goals of professional and ethical review and of formal regulation are to support and to apply the five basic principles which, from the Nuremberg Code (1946) to the WHO and CIOMS International Guidelines, have been regarded as best protecting and benefiting the subject in research: (1) consent of the subject, freely given and after being informed of the risks involved, (2) design of research, shaped by professional expertise within the regulation framework, (3) highest professional standards for researchers, staff,

and facilities, (4) cessation of research in case of unacceptable damage and – a not widely and generally accepted principle – compensation for those damages, and (5) risk-benefit calculation that gives priority to the subject's dignity and his or her right to be free from harm.

In the process of enacting these principles some have proven difficult to establish or do not serve the intended purpose. *Consent* always implies a practical and benefit-oriented cost-benefit calculation of the consenting party. These benefits might be the high appreciation given to risky self-experiments or to those individuals who meet Jonas' moral standards of identification with risky research for the sake of others. Other motives might be large amounts of money, the adventurous and interesting social conditions of the job, or even interest in research itself. For patients it might be the higher degree of medical attention or a higher probability of positive results not otherwise attainable. Some people are incompetent and therefore cannot consent; in order to fit them into the scheme of consent-giving, consent by proxy or by guardians is asked for, but the arguments pro and contra have not yet been settled. *Research design*, as well as result evaluation or change of design, are essential internal requirements within any research undertaking. For any responsible researcher, it is obvious that the subject's dignity and the *primum non nocere* requirement must be considered. Not doing so would be a cause for criminal prosecution. The waiving of right to compensation seems to be immoral and should be declared illegal. Outside reviews of design and of results also are common in other areas of research. Though definitely helpful, such reviews are debatable in cases of unclear shifts in the share of responsibility. *High standards* of expertise and facilities are indispensable prerequisites for good results everywhere. In a regulatory system, however, they mostly will or can only be checked on formal grounds. *Cessation* of experimentation and compensation aspects can best be dealt with on a contractual basis which gives the subject some inalienable rights. Finally, *risk-benefit analysis* is also a basic requirement in any kind of research or otherwise risky enterprise. It is in the best interest of the researcher to minimize risks that might be of moral or of cost relevance. The means by which the five above-mentioned principles are administered include formal regulatory procedures and various forms of review boards, which are responsible for material, technical, and moral reviewing, recommendations, and approval. The *formal regulatory procedures* have been blamed for being wasteful, costly, unnecessary, and for suppressing research or the routine application of scientific results, as well as generally being inflexible and ineffective.[57] *Review Boards*, and especially those established by law, have been criticized for tending to lose their "commitment to the goal of protecting subjects" and instead becoming increasingly "a tool of institutions, to protect themselves, and of researchers, to increase the legitimacy of their activities" ([46], pp. 318–28). It might therefore not be a bad idea to replace I.R.B.s in part by involving insurance companies.

(3) An assessment of criticisms of the various processes of regulation in a growing number of countries might result in different forms of deregulation, the re-installation of the principles of personal responsibility, fair market compensation, and of a rearrangement in regard to the "all or nothing" approval of new drugs or devices by regulatory agencies. Key issues will be: (a) the right to pursue research, (b) the professionalization of research subjects, (c) the differentiation between simply registering by notification and being listed in the pharmacopeia, and further steps of approving and accepting drugs or methods, (d) a reallocation of responsibility between I.R.B.s and individual researchers or physicians, (e) the replacement of experimentation involving persons by increased use of human matter, i.e., tissues, organs, and perhaps whole-brain-death bodies, and (f) most importantly, a professional awareness of responsibilities involved in any medical treatment, especially in New Therapy and in Human Experimentation.

Freedom of research has brought about the incredible progress in all modern sciences, including therapeutic and non-therapeutic knowledge in the biomedical and behavioral sciences. Since morality is only indirectly related to science, it would be improper to control science in order to emancipate morality. The proper approach to giving moral considerations a higher standing in life in general, and in scientific research or innovative therapy in particular, would be the use of educational and cultural means, e.g., by emphasizing the liberal arts education, public moral argumentation, and the example of moral heroes.[58] Since freedom of research and related beneficial results are issues of public policy, and since public policy tends to reflect changes in public or cultural preferences more than interests inherent in the research profession, an effective use of lobbying tools in influencing regulation and legislation may be indispensable for the survival of research in a pluralistic society.[59]

The idea of *professional research subjects* might be entertained more seriously. Given the removal of adverse regulation, proper job descriptions, and adequate compensation and insurance – the proof that an injury is not being caused by experimentation being up to the researcher – such a professionalization could be quite attractive. It has been noted that high salaries, intensive education, daily conferences, high status, choice of experiments, fluctuation of pay in relation to fluctuations in risks, and guarantees against harm would be attractive enough to establish research subject as a professional career ([91], p. 28). A research subject might increasingly play a more active role in research design and evaluation. The idea of self-experimentation, which totally disappeared in the regulation climate, might be rejuvenated in a new form. An educated and experienced research subject might even want to share a more than average risk of the experimentation enterprise in exchange for higher rewards in prestige or cash.

Perfectionist approval regulations might be replaced by less ambitious but more effective forms of regulation: after a rough and expeditious procedure of simple (preliminary) registration, which would not mean approval, drugs or devices would enter a trial period. During this trial period, these methods or drugs would be available to physicians, and others who might need or want them; only after periodically checking the results would the device or drug pass from "unproven efficacy" through different stages of "proven efficacy", simultaneously descending from a category of "high risk" or "unqualified risk" to one of "low risk."[60] Such a reform of registration would allow a fair allocation of risk-benefit to the user-patient and would decrease the amount of other, presently legally required, forms of human experimentation. Assessing the technically complicated and morally not advisable risks of *jeopardizing professional and moral responsibility* could result in a clearer allocation of moral responsibility, scientific accountability, and legal liability, preferably to be carried by the researcher or the research team, which could include research subjects as "subcontractors". Finally, sophisticated progress in methods of biomedical research might *replace persons as research subjects* by increasingly making use of human matter, i.e., tissues, isolated organs, and – if societal opinion is favorable – human bodies with brain death.[61]

Most important, however, is the need to develop, to educate, to refine, to apply, and to protect the burden of *special responsibility, borne by all persons who deal professionally with others*. The medical and behavioral sciences have very particular responsibilities, and these are even higher in New Therapy and in Human Experimentation. According to the *Reichs-Richtlinien*, the medical profession has the *unalienable responsibility to search* for new and better therapy. These responsibilities result in rights that are embedded in their professional and personal responsibility for the life and health of their patients and clients. It is obvious that the moral aspects of the very special unalienable rights and responsibilities given in New Therapy and in Human Experimentation have to be paid special attention in academic teaching of *medical and research ethics* "wherever possible" [90].

It is still an open question to what degree the medical and behavioral occupations are purely scientific, and in how far they are based on a prudent mixture of art, logic, innovation, and experience. The philosophy of regulation so far has not convincingly enough provided a fair and just allocation of risks and benefits and of responsibilities and liabilities of the persons and agencies involved. Yet an awareness of personal responsibility is indispensable in an open and free society based on the flexible balance of liberty and security. We should not forget that the actual state of the medical and behavioral sciences was not achieved as a result of centuries of regulation, but as a result of centuries of relative freedom of thought and intellectual inquiry, self-restricted and self-regulated in the fields of the medical sciences by the professional ethos of the Hippocratic Oath. More than three hundred years ago, Spinoza wrote that such a freedom is possible without harm to beliefs, morals, and peace in society, and that to abolish this freedom would automatically mean the abolition of a peaceable society and of its principles and morals. As Spinoza put it, this freedom

> non tantum salva pietate
> aut rei publica pace posse
> concedi, sed eandem nisi
> cum pace rei publica
> ipsaque pietate tolli
> non posse[62] [107].

Georgetown University
Washington, D.C., U.S.A.

NOTES

* This work and others concerning issues in the field of bioethics were made possible by a grant from Volkswagenwerk Stiftung.

[1] Hobbes' De Cive (1642), Locke's Letter Concerning Toleration (1687), and Spinoza's Tractatus Theologico Politicus (1670), all develop a basic understanding of the dialectical tension and interrelatedness of liberty and security, thus providing the philosophy of contract and covenant as a basis for accepting various forms of Regulation in a Free Society. Veatch [118] deepens and elaborates on these foundations to develop a theory of the role of professional ethics in modern society. See also, Fried [44].

[2] Paternalism, as it appears in different forms in Peer Reviews, Community Review Boards, the Office of Motor Vehicle Registration, the Internal Revenue Service, and other regulatory agencies, has to be understood, according to democratic theory, as being philosophically justified indirectly by legislative authorities, either in accepting existing forms of paternalism, introducing new ones, or accepting the introduction of new ones. Veatch [118] convincingly argues for expanding the classical theory of Social Contract into the areas of contracting between society and highly specialized professions.

[3] First steps for dealing with *res publica – res privata* matters in a contemporary technological society are outlined in regulations and recommendations such as [113], or institutions such as Datenschutzbeauftragter of the Federal Parliament in the Federal Republic of Germany.

[4] See, for example, [72], [115], [82] and [85].

[5] According to Spinoza and others, freedom of thought and of the sciences – *ex definitio* including research – is of the greatest importance for a stable government and an educated and prospering society. Since then, all constitutions of Western states affirm the general freedom of arts and sciences, and condemn their suppression or subjection to political tutelage.

[6] John Locke already recognized the difference between freedom in those areas of life that are of no direct public relevance, e.g., dogmatics, and those of indirect or direct relevance in public affairs, e.g., ethical behavior, acceptance of the legal system, or even ritual slaughtering. Notwithstanding the generally proclaimed "freedom of research", the scientific community so far seems to have accepted public governance of research both by allocation of funds (and therefore the setting of priorities) and by regulations protecting the safety (*securitas*) and dignity (*libertas*) of the individual citizens or the entire society. See, for example, Frederickson [40], who understands such a guidance as necessary and acceptable, and Veatch, who appreciates the individual's "prudent protection" by federal regulation, but also senses the "great risk", especially in "areas normally thought of as the scientific domain" ([116], p.90). Fraenkel [39] describes the increase in regulation in the U.S.A. during the last twenty-five years.

[7] There are numerous examples in the history of science and technology showing that progress has been induced by means other than regulation, including progress that makes previous regulations superfluous. A good example might be the safety of steam engines which finally was achieved by new techniques of welding after decades of exploding boilers and severe casualties in the early nineteenth century.

[8] Most prominent in taking such a restrictive position is Jonas [57], while Veatch [118]

summarily argues pro regulation in extending the Social Contract idea into provider-client relations.

[9] See *Richtlinien* ([90], pp. 174–175), and Sass [101].

[10] The Nuremberg Code, published and translated widely, was first published in *Permissible Medical Experiments, Trials of War Criminals before the Nuremberg Military Tribunals* (October 1946–April 1949); also in [89], pp. 1764–1765. Also quite influential was H. K. Beecher [8] reporting on the lack of ethical awareness at major medical centers.

[11] A first draft was adopted by the World Medical Association in 1962, and was revised in 1964; the final version of the so-called Helsinki-Tokyo Declaration was revised and adopted by the 29th World Medical Assembly in Tokyo in 1975, and was published and translated widely ([89], pp. 1769–1773).

[12] Presented by the British Medical Research Council, Art. 16, 1963 ([89], pp. 1765–1769).

[13] *American Medical Association Proceedings of the House Delegates*, C-66, pp. 189–190 and A-74, pp. 127–131; also in [89], pp. 1773–1774.

[14] See, for example, Schweizerische Akademie der Medizinischen Wissenschaften: *Medizinisch-ethische Richtlinien* ([105], pp. 8–13) and [104].

[15] These Codes are collected in [102], pp. 205–240.

[16] First Guidelines were used internally by the Clinical Center of the U.S. National Institute of Health (NIH) since 1953, by the Public Health Service since 1969, then in 1969 NIH prepared a paper entitled "Protection of the Individual as a Research Subject" which served as a blueprint for the 1971 Federal guidelines, *Institutional Guide to Department of Health, Education, Welfare Policy on Protection of Human Subjects* ([89], pp. 1774–1781). See also [30].

[17] Helsinki-Tokyo Declaration, 1975; I, 1; "The design and performance of each experimental procedure involving human subjects should be clearly formulated in an experimental protocol which should be transmitted to a specially appointed independent committee for consideration, comment, and guidance" ([89], p. 1771).

[18] [113], pp. 30–91; for the earlier Code of Federal Regulations requiring Institutional Review Boards, see [114], and also [92].

[19] [95]; see also, [108], p. 1010, which includes a report, prepared by the British Medical Association's Central Ethical Committee (C.E.C.) in 1980, using information given by 138 local committees of which 101 include representation of the general public and 54 include a general practitioner. The Central Ethical Committee recommended "that the present hospital-based ethical committees should be kept small but have a wider composition, especially including general practitioners". The C.E.C. proposed the following model: two senior hospital doctors nominated by the local executive committee; one junior hospital doctor nominated by the junior medical staff; two general practitioners nominated by the local medical committee; one representative nominated by the appropriate community medical staff; one nurse; and one lay member. "The local committees shall consider and grant ethical approval after the researcher has been given a hearing. Withdrawal of approval shall be reported to an appropriate professional advisory committee."

[20] [12]. These Empfehlungen recommend a Commission of five (two physicians with clinical experience, a medical scientist professionally occupied with research, a lawyer, and a representative of the Ärztekammer, the state or national medical association). The

Commission is to be elected for four years, chaired by one of the physicians, and, on request by a researcher, is to issue a written statement focusing exclusively on ethical and legal considerations, as well as a recommendation achieved by majority vote. Deliberations are confidential, and the applicant also may request a hearing ([104], pp. 617–618). Switzerland will have a fourteen-member Central Medical-Ethical Commission responsible for enforcing the basic principles of the Helsinki-Tokyo Declaration ([104], p. 616). Since 1978, there also are Ethical Committees in four of the Swiss Medizinische Fakultäten, as well as in a number of clinics ([49], [48], [104]). The Medical Research Council of Sweden has established Ethical Committees in all medical schools; the Chairmen of these Committees form a central commission that counsels the Swedish Government. The Deutsche Forschungsgesellschaft [18] appoints Ethical Committees only in very special situations, as ethical considerations always are seen as an integral part of the analysis and realization of biomedical research [30]. Concerning Scotland, see [108].

[21] This is taken from the *Institutional Guide to the Department of Health, Education and Welfare Policy on Protection of Human Subjects*. See also, B. H. Gray [46].

[22] See [103]; among the many reports on institutional ethical committees, see the most recent and informative reports and case studies: [46], pp. 22–25; [83], pp. 19–78; [93], pp. 29–33; [27], pp. 1042–1045; [81], pp. 288–293; [61], pp. 672–681; [109], pp. 718–720; [29], pp. 614–617; [96], pp. 667–670; [119], pp. 168–170; [68], p. 1011. A shift, however, from Peer Review and Commissions (constituted primarily on the grounds of representation and parity) towards Community Reviews seems to be increasingly favored. See, for example, [93], and [29]. *I.R.B. A Review of Human Subject Research*, published by the Hastings Center (since 1979 ten issues yearly), reports and comments on issues related to Institutional Review Boards. Also see [66].

[23] See, for example, [24], and [113].

[24] [33], pp. 2445–2482; see also, [50], pp. 1562–1566, and "Zweite und Dritte Beratung . . . eines Gesetzes zur Neuordnung des Arzneimittelrechts", *Deutscher Bundestag, 7. Wahlperiode, 238. Sitzung, Bonn, 6 May 1976*, pp. 16627–16690.

[25] See, for example, the Proceedings of the Eighth Trans-Disciplinary Symposium on Philosophy and Medicine, [106], especially, W. J. Curran [24]; also Veatch ([116], pp. 75–91).

[26] [126], 49pp. These guidelines follow closely the recommendations of the Helsinki-Tokyo Declaration. Their main purpose is to serve "as a consultative document to Ministries of Health, Medical Research Councils, Medical Faculties, and medical journals as well as any companies" ([126], p. 2). The 1974 *Recommendation on the Status of Scientific Researchers* by UNESCO did not yet reflect or address the Helsinki-Tokyo Declaration. In Par. 14, it speaks in rather general terms about the civic and ethical aspects of scientific research, and, in Par. 29, the protection of the health of the researchers is discussed, including "necessary safety procedures", as well as generally the "monitoring and safeguarding of all persons at risk" ([110], par. 14). The WHO-CIOMS Guidelines definitely will influence ethical considerations in other international organizations, as well as in member states worldwide.

[27] W.J. Curran describes two major approaches: *regulation* on the model of the Federal Drug Administration or *guidance* as provided by the National Institute of Health. Both forms of control presuppose a shared responsibility of administrative bodies, either by funding research or by not prohibiting possibly dangerous goods from being sold [23].

[28] Par. 1: "Es ist Zweck dieses Gesetzes . . . für die Qualität, Wirksamkeit und Unbedenklichkeit der Arzneimittel nach Massgabe der folgenden Vorschriften zu sorgen". Par. 38 requires registration of most drugs according to complicated procedures described in Par. 39; paragraphs 40 and 41 regulate the procedures of controlled clinical tests.

[29] See, for example, the vague formulations as to what are "established and accepted methods" in the DHEW *Guidelines* of 1971, Section E ([85], p. 1777). The *Reichsrichtlinien* based the professional responsibility of the physician on duties that can never be described as exclusively scientific, even though their performance requires scientific knowledge and professional experience.

[30] *Decision* of Verwaltungsgericht Minden, 7, July 1975, 4 K 209/72.

[31] *Decision* of Oberlandsgericht Hamm, 9 January 1964, 9U 145/62.

[32] [64], *cf.* the *Richtlinien* of Europäische Gemeinschaft/European Community (E.C.), 26 January 1965, Articles 5 and 11, I.2.

[33] *Grundgesetz der Bundesrepublik Deutschland*, Art. 5, III, which states that the government may not be partial or biased toward any religious or scientific position or Weltanschauung.

[34] Kriele affirmatively quotes Bundesgerichtshofentscheide (supreme court decisions), *cf. Neue Juristische Wochenschrift*, 1976, p, 380, in stating that "abuse of freedom has been dealt with by firmly punishing such an abuse, not by sacrificing freedom of research in favor of a bureaucratic central administration". In the Federal Republic of Germany only the AMG paragraphs 40–42, and the Strahlenschutzverordnung, par. 41–43, regulate human experimentation with regard to controlled clinical tests and in regard to radiological research. Notwithstanding these areas of special legal regulation, the constitutional framework for Human Research is contained in the Grundgesetz: Art. 1,1 addresses the need to preserve each individual's dignity, Art. 2 notes the constitutional right to life, integrity, and freedom, and Art. 5,3 guarantees the freedom of science and research. Also, civil law as well as penal law provide the principles of lawsuits on the ground of "bodily harm" (Körperverletzung) and "violation of good manners" (*Verstoss gegen die guten Sitten*, St.G.B., par. 226). It has been mentioned that the active regulations of the Arzneimittelgesetz and the Strahlenschutzverordnung might serve as paradigms for additional new regulations in other fields of research using humans as subjects. See, Weissauer [12].

[35] See, for example, *AMG*, Par. 38 and 39, 8, and *Bundesgesetzblatt* (1978), I, pp. 401–402, for regulations regarding notification and labeling of homoeopathica and so-called quack remedies. Requirements for registration of drugs might change depending on cultural preferences or biases, e.g., China in 1930 did not register or import foreign standard drugs if they contained any narcotine (Par. 7) or high doses of poisonous substances (Par. 8). Labeling was not allowed to use obscene or sexually provoking wording or pictures. Also forbidden was wording indirectly recommending abortion or contraceptives, using rhetoric that might be misunderstood, or insulting medicine ([22], p. 3092).

[36] See, for example, [9], pp. 1043–1049; [10]; [69]; [47].

[37] See, [36], pp. 2519–2522; [35], pp. 1094–1096, and [14], pp. 1356–1359. These positions have been criticized by [52], pp. 1087–1153, also [53]. As to some open legal and medical questions in regard to CCT favored by the AMG and liability, see [97], pp. 1182–1187.

[38] A good approach that takes into consideration the legal situation is [13]. Following the

AMG regulation it encourages such a cooperation between general practitioners and research-based pharmaceutical businesses.

[39] "Vorschlag einer E.W.G. Richtlinie über . . Vorschriften und Protokolle für Arzneimittleversuche (Normen und Protokolle)", in *Bundestagsdrücksache* VI-417 (26 Feb. 1970), 3 III 2, understands positive experience of general or clinical practitioners as not sufficient in establishing evidence for efficacy, if not backed up with "scientific evidence". The new E.C. Regulation, 20 May 1975, [32], makes it obligatory for member states to transform into national law the requirements of "controlled clinical trials" (3, I, 2), or "double-blind" control "if possible" (3, I, 3), or accepted statistical standards (3, I, 4), and information about the use of placebos (3, II, 2b2).

[40] With regard to ethical considerations in pharmaceutical research-based companies, see also Cerletti [20] and Pletscher [84]. Ciba-Geigy, the ethical drug company, like other companies, has an independent Ethical Committee for Clinical Research.

[41] It should be mentioned that the AMG, as well as the Deutsche Forschungsgesellschaft, do not have an Ethical Review Board yet, which would include ethicists, philosophers, or theologians, mandatory, as requested by the Helsinki-Tokyo Declaration and the European Community's *Richtlinie*; see, for example, [37].

[42] Baier ([4], pp. 137–148) presents a number of serious reasons for having human experimentation and other forms of research conducted by professional organizations; this would, given adequate funding, decrease regulation and clearly allocate responsibility, whether professional or ethical. [28], p. 42 describes two options for the U.S. legal system: (a) introducing into old case law on battery and negligence new principles such as informed consent, thus creating new case law strongly protecting the subject or patient, or (b) using trust law, and giving the established expertise and responsibility of the trustee priority over the will of the trustor (see, for example, [42]).

[43] That would be true also for research or therapy including placebos. See, for example, [97], pp. 1182–1187. See also Note 34.

[44] Gesetz zur Neuordnung des Arzneimittelrechts, AMG, Paragraghs 40, 1, 8 and 40, 3.

[45] AMG, Paragraphs 84, 2 and 84, 1.

[46] See, for example, [66] and the example given by B. H. Gray ([46], pp. 56f and 85f) in regard to the consent or assent given by subjects and in regard to their capacity for understanding information or the experimental set-up in general.

[47] Also see, [122], pp. 9–14. These Final Regulations allow substantial exemptions for broad fields of behavioral and social science research, ease the requirements for what is called "basic elements of informed consent", while "additional elements of informed consent" may only be required if appropriate, given IRB's discretion for "expedited review" in categories of research "involving no more than minimal risk and for minor changes in research already approved by an IRB" (p. 8366). The proposed changes in the FDA regulations include (a) approval of new drugs "based exclusively on foreign data if the clinical tests are well-conducted and performed by a recognized researcher", (b) acceptance of summaries of case studies instead of full text, which reduces the current 100,000 pp. application substantially, (c) lightening of the internal bureaucratic procedures within the FDA, (d) elimination of approval for minor changes in manufacturing or composition of already approved pharmaceutical compounds, and (e) expanding requirements to report adverse side effects after the drug has been given approval. See also [16], pp. 804–805. Horst Baier has diagnosed the trend to overregulate increasingly as a direct

result of the idea of the Social Welfare State. The citizen's and the patient's right to choose, and to have access to health care and regulated drug production is "an instrument to manipulate the requirements of public health and to dominate the clientele of the social welfare state" ([5], p. 84).

[48] The WHO's *International Guidelines* states that, even though the use of prisoners in biomedical research is not debarred by any of the international codes, the arguments pro and contra are each persuasive on their own grounds so that there would be "no basis for an international recommendation. However, where the "use of prisoners . . . is permissible, this is perhaps deserving of special rules providing for independent monitoring of projects" ([124], p. 13). Case studies and argumentation are found in [60], pp. 1013–1052.

[49] Participation, according to [123], is a fundamental human right, the means of personal development and the opposite of alienation.

[50] Arzneimittelgesetz, paragraphs 40, 4, 1–4 and 41 for parental consent. WHO, International Guidelines ([124], pp. 9–10) states that limitation of child-related research is "axiomatic", parental consent necessary, and cooperation or consent of the child, depending on age, is desirable.

[51] The literature is growing in all related areas; as to children as research subjects, see the debate between Paul Ramsey [86], [87] and Richard McCormick [70] in *The Hastings Center Report*. In his article, McCormick argues that the consent of parents or guardians would be based on a reasonable assumption of what the incompetent or child "ought to want to do for the benefit of society". Federal Regulations for the protection of children and of the institutionalized mentally disabled as subjects in research are still to be issued by the U.S. Department of Health and Human Services. Regulations on research involving the institutionalized mentally disabled were proposed in 1978, *cf.*, *Federal Register*, Vol. 43, No. 53, p. 11328f, and No. 223, p. 53950f, but are still controversial. As to Embryo Research, cf. H.-M. Sass: 1987, 'Moral Dilemmas in Perinatal Medicine', *The Journal of Medicine and Philosophy* **12**, 279–290.

[52] The liberal tradition states *volenti non fit injuria*; see [34], pp. 105–124.

[53] [14], p. 220f accepts liability only in case of fault. As to the lengthy and complicated legal procedures, different in many of the U.S. states, see Levy [67], Miller [71], Childress [21], and Robertson [94]. Childress and Robertson present moral arguments and legal reflections in *The Hastings Center Report*. Robertson insists that if subjects are not compensated for damages, they should at least be fully informed of that fact ([94], p. 30). See also Levine [66].

[54] [16], pp. 650–654 gives some statistics on research-related injuries: among 93,000 subjects in non-therapeutic research there were 0.8 percent reported injured; one person was permanently, 37 temporarily (reaction to drugs, corneal abrasion, electrical burns, assault by other subjects) disabled, 673 had trivial injuries (discomfort, colds, scars, mild allergic reactions); among 39,000 therapeutical research patients the percentage of injured was 10.8 with 43 deaths, 13 permanently disabled, 937 with major, 3253 with minor injuries.

[55] Concerning this, I wish to oppose a judgement prevalent among scholars of continental European philosophy making assessments, as well as technology assessment, a domain of utilitarianism; see, for example, [120], p. 318f. Rabbinic, Calvinist, Lutheran, and Jesuit traditions of procedure assessment, and the classical modern traditions in political and moral philosophy, including even Kant's Categorical Imperative, also are a result of assessing techniques applied preferably to techniques, but also to goals.

[56] See, for example, [3], pp. 389–393.

[57] See the "Final Regulation Amending Basic HHS Policy" [107] as to the proposed easing of wasteful and ineffective FDA regulations.

[58] See, for example, [99], pp. 45–52 and [98], pp. 93–99.

[59] See, for example, [25], pp. 25–26.

[60] The announced deregulation of FDA requirements seems to be one first step in that direction, while the German AMG only is willing to grant exemptions for what some schools of medicine call quack remedies, and which additionally have to be proven to be not harmful. Another negative aspect of administrative regulation of drugs is the possible limitation of available drugs, so called "essential drugs"; the WHO-Essential Drug List (see [125]) might have such a counterproductive effect (*cf.*, [55]).

[61] Philosophical arguments contra are presented by [58], p. 27, calling for a restricted approach to the borderline between life and death, which at least according to many religious and some philosophical positions might not be identical with whole brain death. Arguments pro could be derived from H. T. Engelhardt [31] who gives reasons for differentiating between human persons and human life that is not also personal life. Erwin Deutsch [29] assumes that, if relatives agree, therapy-related research may be permitted by U.S. law for unconscious persons, but no other forms of experimentation are permissible. On the other hand, as already established in Karp v. Cooly ([89], p. 659), everything is permissible for a dying person with his or her consent, even the use of devices such as bloodpumps to replace heart activity regardless of whether they have been tested before, including in animal research. These positions are more or less in line with the Reichsrichtlinien [87], articles 5 and 12d.

[62] See [107].

BIBLIOGRAPHY

1. Abshagen, U.: 1981, 'Ethische Erwägungen. Politische Probleme bei der Arzneimittelprüfung am Menschen. Aus der Sicht der pharmazeutischen Industrie', unpublished paper presented to International Symposium on *Recht und Ethik in der Medizin*.
2. Adams, B. R. and M. Shea-Stonum: 1975, 'Towards a Theory of Control of Medical Experimentation with Human Subjects: The Role of Compensation', *Case Western Reserve Law Review* **25**, 604–648.
3. Arber, W.: 1980, 'Nicht alle Risiken des medizinischen Fortschritts und der biologischen Grundlageforschung sind abschätzbar', *Bulletin der Schweizerischen Akademie der medizinischen Wissenschaften* **36**, 389–393.
4. Baier, H.: 1978, *Medizin im Sozialstaat*, Enke, Stuttgart.
5. Baier, H.: 1981, 'Arzneimittel der Zukunft', in H. Schipperges (ed.), *Müssen Arzneimittel teuer sein?*, Helm, Heppenheim, pp. 77–84.

6. Baier, H.: 1982, 'Gesundheit als öffentliches Gut. Über die Entprivatisierung von Krankheit und Gebrechen im Sozialstaat', *Dreissigste Jahrestagung der Vereinigung Suddeutscher Orthopäden*, Med.-Lit. Verlagsgesellschaft, Ulzen.
7. Bar, Chrand Fischer, G.: 1980, 'Haftung bei der Planung und Foerderung medizinischer Forschungsvorhaben, *Neue Juristische Wochenschrift*, 2734–2740.
8. Beecher, H. K.: 1966, 'Ethics and Clinical Research', *New England Journal of Medicine* **274**, 1354–1360.
9. Bradford, H. A.: 1963, 'Medical Ethics and Controlled Trials', *British Medical Journal* **1**, 1043–1049.
10. Bradford, H. A.: 1967, *Principles of Medical Statistics*, 8th ed.,
11. Branson, R.: 1977, 'Prison Research: National Commission Says "No, Unless . . ."', *Hastings Center Report*, London 7, 15–21.
12. Bundesärztekammer, Fed. Rep. of Germany: 1979, *Empfehlungen zur Einrichtung von Ethik-Kommissionen bei den Ärztekammern*.
13. Bundesärztekammer, Fed. Rep. of Germany: 1978, Prüfung neuer Arzneimittel in der Praxis des neidergelassenen Arztes', *Deutsches Ärzteblatt* **14**, 2773–2774.
14. Burckhardt, R. and G. Kreile: 1978, 'Controlled Clinical Trials and Medical Ethics', *The Lancet* **2**, 1356–1359.
15. Calabresi: 1969, 'Reflections on Medical Experiments in Humans', *Daedalus* **219**, 220–222.
16. Cardon, P. et al.: 1976, 'Injuries to Research Subjects', *New England Journal of Medicine* **295**, 650–654.
17. Carryl, R. C.: 1982, 'Drug Stocks. Ready to Shine', *Value Line. Part II: Selection of Medicine* Value Line, Inc., New York, pp. 804–806.
18. Capron, A. M.: 1975, 'Social Experimentation and the Law', in A. M. Rivlin and P. M. Jimpane (eds.), *Ethical and Legal Issues of Social Experimentation*, The Brookings Institution, Washington, D.C., pp. 127–163.
19. Cavers, D. C.: 1969, 'Legal Control of the Clinical Investigation of Drugs', in P. A. Freund (ed.), *Experimentation with Human Subjects*, Braziller, New York, pp. 225–246.
20. Cerletti, A.: 1979, 'Über Forschung und Entwicklung in der Schweizerischen Privatwirtschaft', *Sandoz Bulletin* **15**, 113–120.
21. Childress, J. F.: 1976, 'Compensating Injured Research Subjects: I. The Moral Argument', *Hastings Center Report* **6**, 21–27.
22. China, Ministry of Welfare: 1930, 'Regulations in Regard to Import and Role of Standard Medicine', *Deutsches Handelsarchiv*, p. 3092.
23. Curran, W. J.: 1969, 'Governmental Regulation of the Use of Human Subjects in Medical Research. The Approach of Two Federal Agencies', in P. A. Freund (ed.), *Experimentation with Human Subjects*, Braziller, New York, pp. 402–454.
24. Curran, W. J.: 1981, 'Clinical Investigations in Developing Countries. Legal and Regulatory Issues', in S. F. Spicker and H. T. Engelhardt (eds.), *The Law-Medicine Relation. A Philosophical Exploration*, D. Reidel Publishing Co., Dordrecht, Holland, pp. 53–74.
25. Delahurst, W. D.: 1976, 'Biomedical Research: A View from the State Legislature', *Hastings Center Report* **6**, 25–26.
26. Dengler, H. J. and F. Gross: 1977, 'Towards a More Rational Regulation of the Development of New Medicine', *European Journal of Clinical Pharmacology* **11**, 233–238.

27. Denham, M. J. *et al*.: 1979, 'Work of a District Ethical Committee', *British Medical Journal* **2**, 1042–1045.
28. Deutsch, E.: 1979, *Das Recht der klinischen Forschung am Menschen*.
29. Deutsch, E.: 1981, 'Ethik Kommissionen für medizinische Versuche am Menschen', *Neue Juristische Wochenschrift* **12**, 614–617.
30. Deutsche Forschungsgemeinschaft: 1978, 'Ethische Aspekte der medizinischen Forschung am Menschen', 19 pp. and appendices, unpublished.
31. Engelhardt, H. T.: 1983, 'Medizinische Technik und Ethische Probleme', in H.-M. Sass and M. Staudinger (eds.), *Wandlung von Verantwortung und Werten in unserer Zeit*, Deutsche UNESCO Kommission, Bonn.
32. European Community: 1975, 'Regulation, May 20, 1975', *Amtsblatt der Europäischen Gemeinschaft*, Nr. L 147 – June 6, 1975.
33. Federal Republic of Germany: 1976, *Gesetz zur Neuordnung des Arzneimittelrechts* (24 August), in *Bundesgesetzblatt* **I**, 110.
34. Feinberg, J.: 1971, 'Legal Paternalism', *Canadian Journal of Philosophy* **1**, 105–124.
35. Fincke, M.: 1977, 'Strafbarkeit des kontrollierten Versuchs beim Wirksamkeitsnachweis neuer Arzneimittel', *Neue Juristische Wochenschrift* **24**, 1094–1096.
36. Fincke, M.: 1978, 'Strafrechtswidrige Methode der klinischen Prüfung', *Deutsches Ärzteblatt* **43**, 2519–2522.
37. Fischer, F. W.: 1979, 'Klinische Forschung und die Ethik der Ärzte', *DFG-Report* I-79; H 14–16.
38. Fischer, F. W.: 1979, '*Medizinische Versuche am Menschen*, Vandenhoeck and Ruprecht, Göttingen.
39. Fraenkel, M. A.: 1975, 'The Development of Policy Guidelines Governing Human Experimentation in the U. S. A Case Study of Public Policy-Making for Science and Technology', *Ethics in Science and Medicine* **2**, 43–49.
40. Frederickson, D. S.: 1978, 'The Public Governance of Science', *Man and Medicine* **3**, 77–82.
41. Freund, P. A.: 1965, 'Ethical Problems in Human Experimentation', *New England Journal of Medicine* **283**, 687–692.
42. Freund, P. A. (ed.): 1969, *Experimentation with Human Subjects*, Braziller, New York.
43. Freund, P. A.: 1969, 'Legal Frameworks for Human Experimentation', in P. A. Freund (ed.), *Experimentation with Human Subjects*, Braziller, New York, pp. 105–115.
44. Fried, C.: 1981, *Contract as Promise*, Harvard University Press, Cambridge, Massachusetts.
45. Gray, B. H.: 1975, *Human Subjects in Medical Experimentation. A Study of Conduct and Regulation of Clinical Research*, Wiley, New York.
46. Gray, B. H.: 1975, 'Assessment of Institutional Review Committees', *Medical Care* **13**, 318–328.
47. Gross, F.: 1979, *Notwendigkeit und Ethik: klinisch-therapeutischer Prüfungen von Arzneimitteln*, Paul Martin Stiftung, Frankfurt.
48. Gsell, O.: 1980, 'Einführung in die Richtlinien der ärztlichen Ethik', *Bulletin der Schweizerischen medizinischen Wissenschaften* **36**, 343–353.
49. Gsell, O.: 1979, 'Medizinisch-ethische Kommissionen der Krankenhäuser in der Schweiz', *Schweizerische Ärztezeitung* **59**, 1345–1358.

50. Hammans, H. (Member of the Subcommittee of the German Bundestag for Youth, Family and Health): 1978, 'Die Arzneimittelzulassung nach dem neuen Arzneimittelgesetz', *Pharmazeutische Zeitung* 123, 1562–1566.
51. Hasskarl, H. and H. Kleinsorge: 1974, *Arzneimittelprufüng-Arzneimittelrecht. Nationale und internationale Bestimmungen und Empfehlungen*, Fisher Verlag, Stuttgart.
52. Hasskarl, H.: 1974, 'Rechtliche Probleme der Arzneimittelprüfung', *Der Informierte Arzt* 2, 1–19.
53. Herken, H. and H. Kerwitz: 1977, 'Der Wirksamkeitsnachweis für Arzneimittel. Basis jeder rationalen Therapie', *Deutsches Ärzteblatt* 74, 2235–2240.
54. Ingelfinger, F.: 1975, 'Legal Hegemony on Medicine', *New England Journal of Medicine* 293, 825–826.
55. International Federation of Pharmaceutical Manufacturers Associations: 1982, *Die Pharmazeutische Industrie* 44, 776–783.
56. Jaffe, L.: 1969, 'Law as a System of Control', in P. A. Freund (ed.), *Experimentation with Human Subjects*, George Braziller, New York, pp. 197–217.
57. Jonas, H.: 1979, *Das Prinzip Verantwortung*, Insel/KNO, Frankfurt.
58. Jonas, H.: 1969, 'Philosophical Reflections on Human Experimentation', in P. A. Freund (ed.), *Experimentation with Human Subjects*, George Braziller, New York, pp. 1–31.
59. Jonsen, A. R. and L. Sagan: 1978, 'Torture and the Ethics of Medicine', *Man and Medicine* 3, 33–49.
60. Katz, J.: 1972, *Experimentation with Human Beings*, Russell Sage Foundation, New York.
61. Kirchhoff, U.: 1980, 'Die Verantwortung muss beim Arzt bleiben. Ethische Probleme in der medizinischen Forschung', *Der Arzt im Krankenhaus* 11, 672–682.
62. Koller, S.: 1977, 'Angriff auf den Fortschritt der Medizin', *Fortschritte der Medizin* 95, 2570–2574.
63. Krause, E.: 1977, *Power and Illness*, Elsevier, New York.
64. Kriele, M.: 1969, 'Stand der medizinischen Wissenschaft als Rechtsbegriff', *Neue Juristische Wochenschrift* 9, 355–368.
65. Lasagna, L.: 1967, 'Special Subjects in Human Experimentation', in P. A. Freund (ed.), *Experimentation with Human Subjects*, Braziller, New York, pp. 262–275.
66. Levine, R. J.: 1981, *Ethics and Regulation of Clinical Research*, Urban & Schwarzenberg, Baltimore and Münich.
67. Levy, C. L.: 1975, *The Human Body and the Law*, Nathan Hershey, New York.
68. Mach, R. S. and O. Gsell: 1979, 'Medizinisch-Ethische Kommissionen und ihre Funktionen', *Schweizerische Ärztezeitung* 59, 1011–1020.
69. Martini, P.: 1953, *Methodenlehre der therapeutisch-klinischen Forschung*, 3rd ed., Springer Verlag, Berlin.
70. McCormick, R.: 1976, 'A Reply to Paul Ramsey – Experimentation in Children: Sharing in Sociality', *Hastings Center Report* 6, 41–46.
71. Miller, R. D.: 1976, *Human Experimentation and the Law*, Aspen Systems Corp., Germantown, Maryland.
72. Mitscherlich, A. and F. Mielke: 1949, *Doctors of Infamy. The Story of the Nazi Medical Crimes*, New York. German version: 1947, *Das Diktat der Menschenverachtung*, Henry Schuman, Heidelberg.

73. National Commission for the Protection of Human Subjects of Biomedical and Behavioral Research: 1978, *Belmont Report: Ethical Principles and Guidelines for Research Involving Human Subjects*, U.S. Government Printing Office, Washington, D.C.
74. National Commission for the Protection of Human Subjects for Biomedical and Behavioral Research: 1977, *Disclosure of Research Information under the Freedom of Information Act*, Bethesda, Md: DHEW (OS) 77–0003.
75. National Commission for the Protection of Human Subjects of Biomedical and Behavioral Research: 1977, *Psychosurgery: Report and Recommendations*, (OS) 77–0001, U.S. Government Printing Office, Washington, D.C.
76. National Commission for the Protection of Human Subjects of Biomedical and Behavioral Research: 1977, *Psychosurgery: Appendix*, (OS)77–0002, U.S. Government Printing Office, Washington, D.C.
77. National Commission for the Protection of Human Subjects of Biomedical and Behavioral Research: 1977, *Research Involving Children*, (OS)77–0004, Department of Health, Education, and Welfare, Bethesda, Maryland.
78. National Commission for the Protection of Human Subjects of Biomedical and Behavioral Research: 1978, *Research Involving Those Institutionalized as Mentally Infirm*, (OS)78–0006 and (OS)78–0007, Department of Health, Education, and Welfare, Bethesda, Maryland.
79. National Commission for the Protection of Human Subjects of Biomedical and Behavioral Research: 1976, *Research Involving Prisoners*, (OS)76–131 and (OS)76–132, Department of Health, Education, and Welfare, Bethesda, Maryland.
80. National Commission for the Protection of Human Subjects of Biomedical and Behavioral Research: 1976, *Research on the Fetus*, (OS)76–127 and (OS)76–128, Department of Health, Education, and Welfare, Bethesda, Maryland.
81. O'Brien, M.: 1980, 'Clinical Research and Its Ethical Control in Durham Between 1974–1979', *Public Health* **94**, 288–293.
82. Ottenberg, P.: 1975, 'Dehumanization and Human Experimentation', in J. C. Schoolar and C. M. Gaitz (eds.), *Research and the Psychiatric Patient*, Brunner/Mazel, New York, pp. 87–103.
83. Pappworth, M. H.: 1978, 'Medical Ethical Committees', *World Medicine*, **13**, 19–78.
84. Pletscher, A.: 1979, 'Medizinische Forschung – wohin?', *Schweizerische Ärztezeitung* **60**, 1567.
85. Powell, J. W.: 1981, 'Japan's Biological Weapons: 1930–1945. A Hidden Chapter in History', *The Bulletin of the Atomic Scientists* **37**, 44–53.
86. Ramsey, P.: 1976, 'A Reply to Richard McCormick – The Enforcement of Morals: Nontherapeutic Research on Children', *Hastings Center Report* **6**, 21–30.
87. Ramsey, P.: 1976, 'Children as Research Subjects – A Reply', *Hastings Center Report* **7**, 40–41.
88. Reating, N.: 1981, 'Federal Regulations Affecting Psychopharmacology Research in the United States', in S. D. Burrows and J. S. Werry (eds.), *Advances in Human Psychopharmacology* **2**, 265–314.
89. Reich W. (ed.): 1978, *Encyclopedia of Bioethics*, Vol. 4, Macmillan Free Press, New York.
90. Reichsgesundheitsamt, Deutsches Reich: 1931, 'Richtlinien für neuartige Heilbe-

handlung und für die Vornahme wissenschaftlicher Versuche am Menschen', *Reichsgesundheitsblatt* **6** (55th Yr.), Berlin.
91. Rist, M. and W. J. Mohan: 1976, 'Wanted: Professional Research Subjects; Rewards Commensurate Risks', *Hastings Center Report* **6**, 28.
92. Robertson, J.: 1978, 'Legal Implications of the Boundaries Between Biomedical Research Involving Human Subjects and the Accepted or Routine Practice of Medicine', *The Belmont Report*, Appendix Vol. 2, DHEW (OS) 78–0014, 16–1 – 16–54.
93. Robertson, J.: 1979, 'Ten Ways to Improve Institutional Review Boards', *Hastings Center Report* **9**, 29–33.
94. Robertson, J.: 1976, 'Compensating Injured Research Subjects: II. The Law', *Hastings Center Report* **6**, 29–31.
95. Royal College of Physicians of London: 1967, 1973, 1981, *Report of Committee on the Supervision of the Ethics of Clinical Investigation in Institutions*, Royal College of Physicians, London.
96. Samson, E.: 'Über Sinn and Unsinn von Ethikkommissionen', *Deutsche Medizinische Wochenschrift* **106**, 667–670.
97. Samson, E.: 1978, 'Zur Strafbarkeit der klinischen Arzneimittelprüfung', *Neue Juristische Wochenschrift* **24**, 1182–1187.
98. Sass, H.-M.: 1980, 'Education as an Indispensable Tool', *Impact of Science on Society* **30**,(2), 437–449.
99. Sass, H.-M.: 1980 'The Quest for Humanism in a Scientific Society', *Zeitschrift für allgemeine Wissenschaftstheorie* **11**, 45–52.
100. Sass, H.-M.: 1983, 'Reichsrundschreiben 1931: Pre-Nuremberg German Regulations concerning New Therapy and Human Experimentation', *The Journal of Medicine and Philosophy* **8**, 99–111.
101. Sass, H.-M.: 1983, 'Standards in Technology and Human Values', in H.-M. Sass and M. Staudinger (eds.), *Wandlungen von Verantwortung und Werten in unserer Zeit*, Deutsche UNESCO Kommission, Bonn, pp. 62–75; 221–236.
102. Schuler, H.: 1980, *Ethische Probleme psychologischer Forschung*, Göttingen.
103. Schultz, Ch.: 1975, 'Rechtliche Zulassung von Versuchen mit Radiopharmaka', *Recht, Technik, Wirtschaft* **8**, 29–41.
104. Schweizerische Akademie der Wissenschaften: 1982, 'Medizinischethische Richtlinien', *Schweizerische Ärztezeitung* **63**, 615–625.
105. Schweizerische Akademie der medizinischen Wissenschaften: 1982, *Medizinischethische Richtlinien*, Schwabe and Co., Basel.
106. Spicker, S. F. and H. T. Engelhardt, Jr. (eds.): 1981, *The Law-Medicine Relation: A Philosophical Exploration*, D. Reidel Publishing Co., Dordrecht, Holland.
107. Spinoza, B.: 1670, *Tractatus Theologica Politicus*.
108. Thompson, J. E., *et al.*: 1981, 'Research Ethical Committees in Scotland', *British Medical Journal* **282**, 1010.
109. *Trials of War Criminals before the Nuremberg Military Tribunals under Control Council Law No. 10: Nuremberg Oct 1946-April 1949*, U.S. Government Printing Office, Washington, D. C., 2 Vols., 1949.
110. UNESCO: 1974, *Recommendation on the Status of Scientific Researchers*, 18th Session of the General Conference, Paris (II B 1).
111. United States, Department of Health and Human Services: 1981, 'Final Regulations

Amending Basic H.H.S. Policy for the Protection of Human Research Subjects', *Federal Register*, Vol. 46, No. 16 (January 26), 8366–8391.
112. United States, Department of Health, Education and Welfare: 1977, *Personal Privacy in an Information Society*, U.S. Privacy Protection Study Commission, Washington. D.C.
113. United States, Senate, Subcommittee on Health of the Committee on Labor and Public Welfare: 1975, *Federal Regulations on Human Experimentation*, U.S. Government Printing Office, Washington, D.C.
114. United States, Senate, Subcommittee on Health, Committee on Labor and Public Welfare: 1973, *Quality of Health Care – Human Experimentation*, 4 Vols., U.S. Government Printing Office, Washington, D.C.
115. United States Senate (Committee on the Judiciary) Subcommittee to Investigate the Administration of the Internal Security Act: 1976, *Humans' Use as Guinea Pigs in the Soviet Union*, U.S. Government Printing Office, Washington, D.C.
116. Veatch, R. M.: 1981 'Federal Regulations of Medicine', in S. F. Spicker and H. T. Engelhardt (eds.), *The Law-Medicine Relation: A Philosophical Exploration*, D. Reidel Publishing Co., Dordrecht, Holland, pp. 75–91.
117. Veatch, R. M.: 1977, 'Hospital Ethics Committee: Is There A Role?', *Hastings Center Report* **7**, 22–25.
118. Veatch, R. M.: 1981, *A Theory of Medical Ethics*, Basic Books, New York.
119. Wagner, H. J.: 1981, 'Aspekte und Aufgaben der medizinischen Ethikkommissionen', *Deutsches Ärzteblatt* **5**, 168–170.
120. Walters, LeRoy: 1976, 'Technology Assessment and Genetics', in R. M. Veatch and R. Branson (eds.), *Ethics and Health Policy*, Ballinger Publishing Co., Cambridge, Mass., pp. 307–329.
121. Weissauer, W.: 1981, 'Ethikkommissionen und Recht', *Bundesärztekammer: Rundschreiben* Nr. J-AK 1/81, dated January 23.
122. Wigodsky, H.: 1981, 'Two Views of the New Research Regulation', *Hastings Center Report* **11**, 12–14.
123. Wojtyła, K.: 1979, *The Acting Person*, D. Reidel Publishing Co. Dordrecht, Holland.
124. World Health Organization: 1975, *Guidelines for Evaluation of Drugs for Use in Man* (WHO Technical Report, Series No. 563), Geneva.
125. World Health Organization: 1978, *Technical Report*, No. 615, Genèva.
126. World Health Organization (WHO) and the Council for International Organization of Medical Sciences (CIOMS): 1982, *International Guidelines for Biomedical Research Involving Human Subjects*, Geneva.

CORINNA DELKESKAMP-HAYES

MORAL APPROPRIATENESS IN HUMAN RESEARCH (A PRACTICAL SUGGESTION AND A THEORETICAL INTERPRETATION)

1

The consent rule is designed to protect the subjects of medical research from various dangers attending their involvement. One aspect of that protection concerns their physical well-being: an adequate balancing of risks and benefits to be determined exclusively by physicians. The other aspect concerns their "well-being" as persons; and here it is disputed by whom balance should be determined. Physicians tend to vote for the investigator himself. They see it as his personal responsibility to disclose the "relevant" information and to avoid any subtle constraints on their subjects' decision. Lawyers, on the contrary, have insisted on greater safety by recommending, among other things, institutional review boards with lay representatives included among their members.

This policy has shown itself to be problematic in two respects. First, it has proven difficult, if not hopeless, to establish general definitions of what "voluntary" and "informed" should mean, in particular when applied to patient-subjects (let alone volunteers, captive populations, children, and so on). It has also remained unclear how the meanings of these terms could be translated into effective rules governing investigators' behavior. Second, even with the conceptual sophistication characterizing research regulations in the United States today, Barber [1, 2, 3,], Gray [13, 14, 15], and others have shown that patient-subjects' opportunity for rational self-determination has not proportionally increased. Thus the existing regulations are not sufficiently effective in accomplishing their goal.

In emphasizing the latter problem, I am operating under certain presuppositions. First, medical research, even of little or no therapeutic value for patients, must continue [8]. Second, I agree with Sissela Bok in concerning myself neither with the clearly trivial, nor with the clearly unfavorable, but only with the intermediate cases of medical research, where consent is appropriate and necessary. Third, I agree with Jonas [20], and many others in considering the relation between physicians and patients to be essentially a personal, and thus a moral one.

S. F. Spicker, I. Alon, A. de Vries, and H. T. Engelhardt (eds.)
The Use of Human Beings in Research. 91–101.
© 1988 *by Kluwer Academic Publishers.*

Therefore, the matter of patient research and of consent is not only – as Professor Sass has emphasized – one of risk-taking but also one of rightness.

Consider the analogy of education. The amount of influence exerted by parents and teachers on unconsenting or resisting individuals is indeed frightening. An extenuating circumstance, however, lies in the fact that those individuals will eventually be in a position to pass judgement on the manner in which they were treated. Good parents and teachers make sure that in view of such retrospective judgement their educational experiments can be *endorsed* by their matured subjects.

Returning to the issue of this conference, my practical suggestion is to institutionalize a similar concern for endorsability into medical research. For the disputed question mentioned above, this implies that the only one who is in a position to evaluate – not of course the physical aspect of his treatment as a research subject but – the humane aspect of that treatment, is the subject himself. While the existing consent requirements will give him some general idea of what to expect, the full implications of his cooperation will become clear only as the matter proceeds. (For similar ideas see [4], [15], [21].)

Regardless of what he is told about alternative courses of action open to him, a patient will usually and should indeed be permitted to trust his physician. Afterwards, however, he will decide if that trust was justified.

Suppose that a routine interview, focussing perhaps only on a random sample (and obviously, this pertains to non-patient volunteers as well) conducted by some independent observer, perhaps even a representative of the funding agency, would take place after a subject's involvement has ceased. Suppose it would question the subject as to the adequacy of information received about the project itself as well as its bothersome or painful implications, as to the degree to which his cooperation was appreciated, and whether he would recommend such an experience to others. Suppose, furthermore, the results of such humanistic auditing would be translated into some sort of investigator-evaluation by subjects – an institutionalized substitute, so to speak, for the kind of talking back that educators will get from their grownup children. And suppose, finally, that funding agencies would feel obliged to consider the result of such evaluations in view of the future allotment of research grants – not only to that particular investigator, but also to his institution, as being responsible for maintaining a properly moral atmosphere among its members. What would the results of such a policy be?

I should like to speculate that they would operate on two levels. For the institutions themselves it would finally become a matter of vital self-interest to ensure that research is not only appropriately started but equally so continued and terminated. The physician-investigators, on the other hand, would know that their future standing as researchers depended not only on their scientific, but also on that human performance, which is required in view of the unique vulnerability of their "research material". He would have to become rather careful in keeping his two roles distinct in the eyes of his patient. The decision, how much information should be disclosed in order to leave the subject satisfied in the end and how much effort should be spent in order to relieve as many subtle constraints as may be operative in each individual case, would be a matter exclusively between investigator and subject. Nor would it all have to happen at the start of their encounter (see the problems discussed in [11]). Investigators would have to feel motivated to keep in touch with their subjects' upcoming questions and worries. Researchers' present tendency of using consent for shunning responsibility [14] could be reversed. Instead, self-interest would move them to satisfy their subjects' curiosity on how the experiments were progressing. Indeed, researchers might come to further that very understanding and ability to identify with the experimental task in their patients, which Jonas has all too idealistically recommended as an initial criterion for subject recruitment.

Thus, while the original consent of patients can rarely be termed rational or voluntary in a strict sense, it is possible, I think, to eventually transform their initial unfreedom into a "freedom *ex post*", or to remedy any unfairness of the consent-situation by giving due weight to the subjects' final endorsement.

This means asking a lot of busy researchers. But by adopting a policy that includes such an *ENDORSEMENT RULE* a considerable amount of protective red tape might be avoided [5]. Fear of legal consequences [22] would no longer have to impede medical progress [16]. The public trust of medical researchers could be regained, which would, in turn, render subject recruiting easier. And anyway: while researchers and subjects jointly serve the well-being of future patients, researchers also serve their own future careers ([13], p. 226). It is proper then that they (just as the institutions serving their own academic reputations) should take special efforts in order to make up for that moral imbalance. In this context, just as with rendering education palatable, I see nothing wrong with a certain amount of bribery – such as characterized, for example, the metabolic ward described by René Fox. If tax money should be expended in order to render the research wards a trifle more comfortable and just slightly better staffed than the rest of our hospitals – would that not be a decent way of sharing the load of research with those who suffer from it?

So much for the practical suggestion. Its presuppositions and implications have only been roughly sketched. But their elaboration can be left to whoever may find this worthwhile.

2

What are the conditions under which such a suggestion could be rendered acceptable to physicians? Three sources of resistance come to mind, and they concern three aspects of the suggested procedure. The first is ethical, the second relates to political theory, and the third to prudential aspects of policy-making. In discussing these sources of resistance, the promised theoretical interpretation will concern a Hegelian view of medical research and practice.

The first source of resistance lies in the behavioristic arguments commonly used to defend the "regulative philosophy", as Professor Sass has nicely called it. Regulations are often recommended by those who view human behavior as a matter, not of personal convictions or previous training, but as conditioned by the actual working environment. Behavioral changes are effected in such a view, not on the basis of persuasion or information, but only by means of institutionalized pressures that will channel a person's self-interest in the desired direction ([1], p. 188). In defense of such an interpretation, Eliot Freidson has argued that there is little empirical evidence that the practice of physicians is influenced by moral considerations. Just as with everyone else, monetary benefits, personal customs, and human laziness play a leading role [10]. With regard to research, Gray ([13], pp. 70, 249) and others have pointed to the pressures attending academic survival or ascendence within hierarchical institutions. They could prove these pressures to be the predisposing conditions for dehumanizing experiments.

Understandably, sociological findings of this sort, however undeniable in themselves, would leave the subjects of such behavioral investigation somewhat dissatisfied ([14], p. 45). After all, the suspicions raised against physicians could be returned on an equal basis: lawyers, sociologists, and philosophers, just as physicians, claim to be moved by a humanistic concern for innocent victims – in the one case, of medical malinvestigation, in the other, of hitherto untreatable diseases. But lawyers, sociologists, and philosophers, as well as medical researchers, welcome the opportunity for publishing papers ([19], p. 264). And just as the latter are exposed to the temptation of implicitly denying the human dignity of their research-subjects, so are the former, insofar as they evaluate *their* subjects on the basis of an essentially reductionist model. The moral excellence of a person or of an act is not a quality of the sort that could be empirically asserted or denied. It arises – in Hegelian terms – from a relation of mutual acknowledgement ([17], pp. 445–450). Viewing physicians exclusively in terms of the regulatability of their behavior means shunning the issue of their possible acknowledgeability, and in this sense indeed denying their personhood. But how could we, in disrespecting physicians, entice them to respect their patients?

The resistance of physicians against outside regulation of medical procedures arises in part from an implicit sensitivity to that moral inappropriateness. An explicit analysis of that inappropriateness is found in Hegel's *Phänomenologie*. This sketch is of course not on the philosophical foundations of medical ethics. I shall therefore have to restrict myself to hints that will be sufficient for the Hegel connoisseur, and to short paraphrases that will render the significance of Hegel's notions comprehensible – so I hope – to non-philosophers.

The physician's – and along with it the medical researcher's – ethos has traditionally been linked to special duties (for the individual) as well as special privileges (for the profession). The former are circumscribed by doctors' conscience which obligates them to responsible and altruistic care. The latter arise from a corresponding distance with regard to public interest and established law. Such exclusive insistence on individual conscientiousness also characterizes Hegel's "seiner selbst gewissen Geist". Hegel's term points to a stage in the self-realization process of mind which is vulnerable to the double reproach of evil, motivated by that latter distance, and of hypocrisy, motivated by that former claim of conscientiousness ([17], pp. 464 ff). This latter reproach amounts to the very reduction of avowed idealism to narrow self-interest, to which doctors are often subjected ([17], p. 467).

But Hegel also repudiates such criticism. He observes the depravity – "das niederträchtige Bewußtsein" – of those who, while keeping themselves clean of active involvements and shunning the responsibility of dutiful commitment, have an easy task with criticizing ([17], p. 468).

Finally, Hegel's solution to that quarrel is also applicable to the conflict between medicine's claimed ethos and the quest for public regulation. In his view that quarrel is overcome as soon as the insight into its own one-sidedness, elicited in the party criticized, is matched by a corresponding confession on the side of the critic. The breaking of the hardened heart, in his terms, represents the transition to the desired state of mutual ethical acknowledgement. Thus insofar as medical resistance to outside regulation is motivated by a sensitivity to the depersonalizing implications of the quest for such regulation, a more reflective, more self-critical way of motivating that quest would represent a similar transition.

What would be the consequences of such a change in attitudes, and how would it affect that very quest? An answer to this question will emerge from a consideration of the second source of medical resistance.

Ever since constituting itself as a profession in the latter half of the 19th century, medicine has claimed, and has been granted, a special status within society. It has been accorded considerable autonomy, and even a dominance over related professions [10]. This special status and the arguments justifying it are commonly cited by physicians' claiming the pointlessness and even detrimental consequences of bureaucratic meddling with – for example – medical research. All responsibility must rest with the individual physician-investigator, or at most with his professional peers [23]. Formal rules instituting public accountability, by undermining physicians' self-respect and by exposing them to legal risks, will weaken that sense of responsibility. In the end, the patients will be worse off than before (see for example [8], pp. 212, 224).

Such claims to professional sanctity have elicited a number of sociological demythologizations. Freidson ([10]) argues that the special status of medicine in society cannot be rationally defended on the basis of the supposedly unique properties of medical knowledge and practice. The special privileges of the medical profession should be abolished; research and practice must be subject to extensive public scrutiny [13].

In view of such iconoclastic leanings, the medical hostility to any infringements on the integrity of medicine's concerns is easily explained. Nevertheless, it is possible that quest for a moderate amount of research regulations can be reconciled – again – with a more reflective respect for professional integrity. We only need to consult Hegel's analysis of the sources of public ethos in his *Rechtsphilosophie*.

In the *Phänomenologie*, the ethical implications of action were considered in terms of mind's development from embeddedness in social decorum to its achieving self-conscious morality. The philosophy of law, however, systematizes the social, legal, and political presuppositions for that very relation of mutual acknowledgement, which here distinguishes such decorum as a more developed stage from what is now denounced as "mere" morality. Decorum, "Sittlichkeit" (or propriety, "Rechtschaffenheit") is achieved when people have learnt to "naturally", out of their own accord, incline towards what is generally acknowledged as being required, i.e., what is set down in the laws of a well-governed state ([18] § 150, pp. 160 ff).

That "own accord" linking the individuals' conscience to those laws is effected by the more immediate, narrow groupings into which – on the basis of their birth or their occupation – the citizens form themselves: the "Stände", and "Korporationen", reminiscent of medieval guilds ([18], §§201, 252, 301). By identifying themselves with such groups, as with the immediate sources of individual well-being, security, and respectful acknowledgement, those citizens also commit themselves to the general welfare of the state, the protector of such groups.

Of course, Hegel does not deal with a medical profession, since it did not exist in his time. But when it did establish itself in Germany, it adopted the notion of "Stand" in a Hegelian sense, emphasizing its autonomy, its commitment to the state, and the particular ethical dimensions of medicine's activities. Viewing the medical profession in terms of such "ständische Sittlichkeit" thus amounts to the desired acknowledgement of its ethos. This view permits both to reject non-physicians' quest for very far-reaching public controls and to dispel physicians' exaggerated fear of such controls.

First: from a Hegelian standpoint professional autonomy is not dependent on the sort of scientific or even rational justifications Freidson has denied to medicine. Rather, it provides the natural intermediate links by which individual citizens may come to have a regard for the state as a whole. Hegel, in criticizing the tendency of his own time to dissolve the corporations, even insists on the necessity to let those communities rule their own concerns in whatever arbitrary and unreasonable ways they may see fit ([18], §289).

Second, however, such groups are placed under the explicit purview of the state, as the guarantor of the general welfare ([18] §§230 ff). As medical research presents opportunities for violating individuals' rights for the sake of medicine's own parochial interests – applying Hegel's view – the state must interfere.

Thus, that very interpretation of medicine in terms of "Sittlichkeit", which was designed to take the moral sting out of the quest for regulations, now appears to be inseparably tied up with a view of the medical profession that renders such regulations, provided they are discerningly employed, a matter of course.

The difficulty of such discernment presents the third source of medicine's resistance to regulation.

Physicians may even grant that, in acting as investigators, they occupy a position comparable to that of other professions. They may even concede that this particular aspect of their activities calls for outside scrutiny. After all, physicians themselves were the first to oppose the abuses they had observed in medical research ([6], p. 309). Nevertheless, these same physicians may well reason that once the public has been granted the competence to supervise a very restricted area, such supervision will inevitably, by the sheer momentum of bureaucratic involvement, be extended to other areas as well. Indeed, would not just such an endorsement rule, as suggested in the first part of this paper, gradually tend to be extended into the humane aspects of ordinary medical practice as well? Or how could medical research, which in its more or less implicit forms is inseparable from most medical practice, be at the same time so principally distinguished from the latter, that such gradual extension could be guarded against?

In view of this third source of resistance the "special professional status" of medicine must now be scrutinized.

Let me begin by specifying my analogy between the manipulative potential inherent in the educational and the medical research situation. Education is indeed like medical research in that concern for future subject endorseability will render it more humane. Education is on the other hand, unlike medical research and like general medical practice in that any form of institutionalized subject evaluation will eventually prove counterproductive. Insofar as the results of student evaluation of faculty members' performance are taken seriously by department heads, professors will come to adjust their performance to the students' stated present needs, as opposed to their as yet undeveloped future needs.

They will also become less inclined to take risks, as for example the risk of investing one's person. While in education this may be a price that one is willing to pay for the sake of greater overall teaching effectiveness, medical practice, if it is indeed at bottom a personal or moral matter, would indeed – as was claimed by many writers – change for the worse.

But how could such a personalist interpretation fit in with Hegel's view of a profession? This view has rested on the possibility of acknowledging the ethical implications of medicine in terms of Hegel's "Sittlichkeit". On closer scrutiny, however, those terms – while being undeniably relevant – are clearly not sufficient. "Sittlichkeit" hinges on the relation between the individual and the state. Yet medicine has always primarily committed itself to individual patients as such; and physicians have considered themselves their special advocates even in opposition to the claims of the state. Indeed, physicians have defended the necessity of violating laws for the sake of medical responsibility. Hegel of course pronounces such an attitude to be principally invalid. However, he is not primarily concerned with states as they really exist. His philosophical reconstruction rests on the hypothesis of an ideal well-governed state in the Platonic sense of the term. Hence, if we want to reconcile the liberties physicians sometimes – and with due discretion – feel justified in taking with our Hegelian account, we have to assume that physicians operate in an area to which even the ideal state's perfection does not extend. Such a possibility is at least implicitly suggested by Hegel.

Poverty and generally all misery that may come on individuals are cited as the one unavoidable source of that imperfection ([18], §242). Regardless of any institutionalized measures to relieve human suffering, there will nevertheless always remain sufficient opportunity for spontaneous individual assistance. ("Mildtätigkeit"). Hegel lists the provision of hospitals on the side of the state. Nothing therefore hinders us in adding to the individual's side that sympathy which practicing physicians, over and above the call of duty, may invest in their patients. Hence, a medical profession, understood in Hegelian terms, must occupy a place not quite covered by Hegel's "official" doctrine.

This doctrine assigns personal and emotional concerns to the sphere of mere morality. Morality's dialectical immaturity lies in its assuming an opposition between the supposedly oppressive forces of reality and the supposedly oppressed humanity of individuals; or, as Hegel later describes it, on the standpoint of morality the conditions for the self-realization of self-consciousness are not yet conceived in a properly dialectical or mediated way. This misconception is corrected only by mind's advancing to the stage of "Sittlichkeit." Only here has it become evident that the individual must already have identified itself with its now acknowledged public duties, in order to be able to realize its now matured self within the legal framework.

However, Hegel's diagnosis of morality's immaturity is valid only insofar as an ideal state takes care of all citizens' needs. Insofar as diseases represent an aspect of citizens' potential needs that cannot be sufficiently relieved by the state, a certain portion of moral mindedness among those in charge of diseases is appropriate. It follows that medicine has a twofold ethical nature, one (covered by the contract model) pertaining to its essentially professional "Sittlichkeit", the other (covered by the fiduciary model, [12]) to the personal element of morality which is inseparably linked with medical practice. Hence, a specifically personal ingredient is indeed reconcilable with an expanded Hegelian interpretation of the medical profession.

The third source of medical resistance to regulation arose from a difficulty of conceptually safeguarding against the feared transition from regulating medical research to regulating medical practice. This conceptual difficulty has now disappeared. The nature of medical research is exhausted by the "professional" analysis of medicine. Its ethical implications arise from the need to comply with what is publicly required. The distribution of intra-professional and external control mechanisms is therefore a matter of political prudence. The nature of medical practice, however, is not exhausted by the "professional analysis". Its ethical implications also comprise a moral concern over the patient's well-being. Its unavoidable public control – as where individual rights are at stake – should therefore be restricted in such a way as to leave the humane involvement of physicians untouched. Any denied changes in the latter can only be gently suggested by duly respectful moral appeals.

As a result, I hope to have shown the theoretical ways in which the practical suggestion introduced above could be rendered acceptable to physicians.

Freigericht 1, Somborn
Federal Republic of Germany

BIBLIOGRAPHY

1. Barber, B. et al.: 1973, *Research on Human Subjects*, Russell Sage Foundation, New York.
2. Barber, B.: 1976, 'The Ethics of Experimentation', *Scientific American* **234**, p. 25.
3. Barber, B.: 1980, *Informed Consent*, Rutgers University Press, New Brunswick, New Jersey.
4. Barthlomé, W.: 1977, *The Ethics of Nontherapeutical Clinical Research*, U.S. Government Printing Office, Washington D.C.
5. Broad, W. J.: 1979, 'Proposal for Ethics Boards', *Science* **205**, p. 285.
6. Donagan, A: 1977, 'Informed Consent', *Journal of Medicine and Philosophy* **2**, p. 307.
7. Eilenberg, M. D.: 1973, Contribution to Discussion in 'Research Investigations in Adults', *British Medical Journal* **2**, p. 210.
8. Eisenberg, L.: 1977, 'The Social Imperatives', *Science* **198**, p. 1105.
9. Fox, R.: 1959, *Metabolic Ward*, Free Press, Glencoe, Illinois.
10. Freidson, E.: 1970, *The Profession of Medicine*, Dodd, Mead, New York.
11. Fries, J. F. and Loftus, E. F.: 1979, 'Informed Consent', *Science* **204**, p. 643.
12. Goldiamond, I.: 1978, 'On the Usefulness of Intent', *The Belmont Report* Vol. 2, DHEW Publ. No OS–78–0014.
13. Gray, B. H.: 1975, *Human Subjects*, John Wiley, New York.
14. Gray, B. H.: 1978, 'Informed Consent', *Annals of the American Academy of Political and Social Science* **435**, p. 38.
15. Gray, B. H. et al.: 1978, 'Research Involving Human Subjects', *Science* **201**, p. 1094.
16. Hill, A. B.: 1963, 'Medical Ethics', *British Medical Journal* **2**, p. 1043.
17. Hegel, G. W. F.: 1952 (1807), *Phänomenologie des Geistes*, Felix Meiner, Hamburg.
18. Hegel, G. W. F.: 1821, *Grundlinien der Philosophie des Rechts*, Nicolaische Buchhandlung, Berlin.
19. Ingelfinger, F. J.: 1975, 'The Unethical in Medical Ethics', *Annals of Internal Medicine* **83**, p. 264.
20. Jonas, H.: 1970, 'Philosophical Reflections', in P. A. Freund, (ed.), *Experimentation with Human Subjects*, George Braziller, New York.
21. Marsh, F. H.: 1978, 'An Ethical Approach', *Ethics in Science and Medicine* **4**, p. 135.
22. (Editorial): 1967, 'The Relation of the Clinical Investigator to the Patient', *American Journal of Cardiology* **19**, p. 892.
23. Williams, G. E. O.: 1970, Contribution to Discussion in 'Medical Ethics and Society', *Public Health* **85**, p. 10.

AMOS SHAPIRA

PUBLIC CONTROL OVER BIOMEDICAL EXPERIMENTS INVOLVING HUMAN BEINGS: AN ISRAELI PERSPECTIVE

A. THE CASE FOR MULTI-DISCIPLINARY PUBLIC CONTROL OVER BIOMEDICAL HUMAN RESEARCH

Current dramatic developments in biomedical research on humans have produced a conflict between two weighty social concerns: On the one hand, the interest in the advancement of biomedical science for the benefit of the present and future generations;[1] and on the other hand, the principle calling for treating the individual as an end in itself, with his rights to privacy, autonomy, and self-determination adequately protected. These two all-important social concerns require proper consideration and mutual balancing.

One cannot escape the feeling that, until recently, the principle of individual privacy, autonomy, and self-determination has not been accorded its due consideration. The main concern of human rights champions has focused on safeguarding the individual from overreaching political organs of government. Hence close attention has been given to the freedom of expression, association, occupation, and the other conventional civil and political liberties. Only marginal regard has been paid to the need for reinforcing the decision-making potency of the individual *vis-à-vis* other concentrations of power in society, notably the resourceful, paternalistic, often overbearing medical profession. Present-day consumeristic trends extend more and more to the sphere of patient/subject – physician/investigator relations, with a view to enhancing the willingness and capability of medical clients to affect more actively professional decision-making concerning their well-being.

The process of biomedical experimentation on human beings entails various complex issues, such as:

(a) the distinction between medical "research" and "therapy";
(b) the assessment of the scientific validity and social utility of research protocols;
(c) the risk-benefit calculus, namely, balancing possible risks to particular research subjects with the potential benefit to them and/or to scientific progress;

(d) the "informed consent" of the subject as a prerequisite to involving him in clinical trials;
(e) the criteria and procedures for recruiting subjects for biomedical experiments, with special concern for particularly vulnerable populations such as children, prisoners, the mentally sick, the chronically ill, the very old and the poor;
(f) insurance and fault or no-fault compensation of subjects injured in the course of experimental research.

From an overall social vantage point, a most important consideration is that of devising a viable mechanism of control over all major phases of the biomedical research process – research policy formulation, research project administration, and research consequences evaluation. The first supervisory method which comes to mind is some form of regulation exercised by the investigator himself within the concrete framework of his special relationship with the subject. Naturally, self-regulation is the model traditionally preferred by physician-investigators. Such preference springs not only from parochial self-interest, but is also deeply entrenched in a professional conviction that, in the final analysis, the only real guarantee for the protection of research subjects lies in the professional conscience, integrity, and judgment of the individual investigator relying on his professional training and experience. Men of medicine habitually reiterate, in this connection, the classical maxim of "primum non nocere" ("above all, do not cause harm"). Yet, the very essence and scope of the notion "harm" lend themselves to varied interpretations. Evidently, not all physician-investigators subscribe to the view that "harm" may encompass instances of undue interference with the freedom of choice of individuals who become subjects in clinical trials without their full knowledge and voluntary consent. The truth of the matter is that individual physician-investigators cannot be expected to reckon with and adequately accommodate all the complex values and interests intertwined in the human experimentation field. Medical education and clinical experience do not, in themselves, justify entrusting the exclusive decision-making power in these matters to the sole discretion of medical practitioners and researchers.

A more structured control system is that operated by the scientific community through such organs as the managing authorities of hospitals, medical school boards, scientific organizations, professional associations, and editorial boards of scientific journals. At present, scientific community control is entrusted mainly to "peer review committees" or "institutional review boards", namely, institutional bodies specially designed for supervising biomedical human experiments conducted in hospitals, universities, and other scientific centers. This kind of control is currently widespread within the scientific community. It has virtually superceded past forms of naked self-regulation exercised by individual physician-investigators. The establishment of institutionalized modalities of peer review attests to a growing recognition by the scientific community that a reasonable supervisory arrangement should be maintained in this field. Institutional review is also expected to forestall outside interference by laymen in the human experimentation process. However, one must admit that the professional community has not been very successful in constructing, of its own initiative and without outside pressure, an effective control mechanism. By the same token, it has failed to prevent serious violations of human rights – particularly rights of children, prisoners and sick people – perpetrated in the name of scientific sanctity and freedom of research. It seems that outside intervention is not only necessary to protect the legitimate interests of patient-subjects, but is also instrumental in accelerating much desired self-criticism thus serving, in the end, the true objectives of the scientific community itself.

A third control model builds on the concept of public supervision exercised by some external regulatory mechanism of a centralized (governmental, for instance) nature. Such a public control system is usually designed to augment internal institutional review processes. It might be managed on an *ad hoc* basis – e.g., through the mass communications media, public symposia, pressure and interest groups, and commissions of enquiry. It can also be operated in an institutionalized way by, for instance, a governmental agency acting under law. A central supervisory agency of this kind could be authorized to lay down, through primary or secondary legislation, the basic policy guidelines, ethical yardsticks, and administrative procedures governing biomedical research on humans. It can also be empowered to supervise regularly the performance of institutional review boards, including exercising central control over the institutional processes of obtaining the "informed consent" of subjects. The involvement of a public control mechanism in the human experimentation field is likely to serve the purposes of formulating a sound and consistent biomedical human research policy, according appropriate weight to various societal interests and values, promoting professional and public discussion of the underlying issues, and thus generally enhancing public awareness of the social significance of scientific research.

It is understandable that many physician-investigators are inclined to shun the very notion of outside interference with their professional research activity. The involvement of ethicists and jurists in the human experimentation decision-making process is instinctively deemed by the biomedical research community as an improper meddling in its own legitimate domain. Yet, to paraphrase a famous cliché about war and generals, biomedical research is too important to be left to the sole discretion of scientists. There is no denying the fact that the scientific community – even in societies which are essentially humane, open and democratic – has failed, time and again, to prevent instances of over-reaching, aberration, and even occasional outright abuse in biomedical experimentation involving human beings [11]. The field of human research is complex and baffling. It entails trans-scientific dimensions – moral, ethical, socio-economic, and legal. It presents societal dilemmas and therefore society at large is entitled, and obligated, to partake in fashioning balanced solutions to those intricate dilemmas. Human research as a total societal affair naturally calls for multi-disciplinary treatment.

To be sure, men of science are, and should always be, the prime activators of biomedical research on humans. Their primary responsibility is to elaborate scientifically sound research proposals and to carry them out in a professionally valid manner. Philosophers should enrich the human experimentation decision-making process with relevant extra-scientific humane and social values. Their task is to elucidate and articulate the olympian moral and ethical principles which speak to the ever-existing dilemmas of individual rights and social responsibility, of self-determination and professional privilege, of present individual sacrifice and future societal progress. Men of law should then be called on to mediate between specific research proposals designed by scientists and the over-arching tenets illuminated by philosophers. Lawyers are best suited, by schooling and training, to reason from broad norms to specific fact situations, to analyze concrete cases in the light of normative abstractions, to synthesize guiding rules from an aggregation of particular rulings, to balance varied – and sometimes contending – claims of interest and fairness. Multi-disciplinary cooperation can thus provide the best available framework for analyzing and solving the multi-faceted problems entailed in the complex field of biomedical experimentation with human beings.

Obviously, the appropriate system of public control is a direct function of the profile of the community in question and should reflect society's particular value system and socio-political realities.[2] Thus, for instance, the prevailing general attitude to the notion of close central governmental regulation in socio-economic matters, and to the constitutional precepts of individual liberty, privacy, and autonomy, is bound to inform and shape the particular human experimentation review mechanism most suitable to the community in question. Its very size is also a factor with which to reckon. In a small country like Israel, where members of the scientific community frequently know each other personally, peer review alone is likely to amount in effect to no more than self-regulation.

B. THE GENESIS OF PUBLIC CONTROL OVER BIOMEDICAL HUMAN RESEARCH IN ISRAEL

Until recently, Israeli law did not prescribe a specific scheme for prior, preventive review of biomedical experimentation involving human beings. The licensing of drugs in Israel is generally regulated under the Pharmacists Regulations of 1986 ([25], p. 906). The Pharmacists Regulations prohibit the manufacture, marketing, importation, and authorization for use of a pharmaceutical, unless it is registered under the Regulations and according to the registered specifications. The Director-General of the Ministry of Health (hereafter Director-General), who is in charge of keeping the Pharmaceuticals Registry, is duty bound under the Regulations not to register a drug unless, among several other pre-conditions, he is satisfied that the drug in question "is not harmful or likely to harm health and is effective for the purpose prescribed to it" ([25], 6 (2)(2)). Both freedom from risk and functional effectiveness are thus made a *sine qua non* to the licensing of a drug in Israel. These requirements seem reasonable in view of the fact that registration authorizes the mass manufacture, marketing, importation, and use of a drug in the free market.

Certain kinds of pharmaceuticals are exempt from the requirements set forth by the Regulations. Among them is a drug "intended for medical or other scientific research, with the authorization of the Director-General"[3] ([25], Reg. 29(a)(5)). The apparent rationale underlying this exemption is that a pharmaceutical designed for medical research, which by its very nature is limited in scope and likely to be controlled, need not satisfy all the strict criteria governing the widespread administration of licensed drugs. In a recent decision rendered by the Israeli Supreme Court ([20], p. 477), the view has been expressed that this exemption may only pertain to the application of an unlicensed drug for the purposes of pure medical research, as distinguished from therapeutical usage. With due respect, the present writer does not subscribe to this view. Neither the wording of the regulation in question nor its supporting rationale mandates a narrow reading of the expression "medical or other scientific research" so as to confine it to pure research only. In my opinion, the Director-General may exercise his discretion under this regulation to authorize – with all prerequisites he may deem fit to prescribe – the administration of a drug for research purposes in the context of therapeutical treatment as well.

In the absence of specific legal arrangements for the regulation of biomedical experiments on humans, the general Assault and Negligence (Malpractice) principles of Israeli tort and criminal law have provided the main legal foundation on which an allegedly injured patient-subject could possibly base a cause of action against offending physician-investigators [3] [7]. Israeli tort and criminal law imposes on a physician a general obligation to obtain the patient's prior consent to the proposed medical treatment. A medical intervention executed without the patient's consent is regarded ordinarily as an act of assault, namely, an intentional touching of the body of a person without his consent. The mere act of unauthorized touching may constitute an actionable tort, or an indictable offense, even where no actual bodily harm resulted from the touching in the circumstances. A medical procedure undertaken without the patient's consent, as well as treatment administered in a manner substantially different from what the patient had consented to, could render a physician liable for the tort or offense of Assault. The same principle is generally applicable in the context of clinical trials. The law, nonetheless, carves certain exceptions and provides various defenses to an alleged act of medical assault. Furthermore, judges seem instinctively reluctant to attach the "assault" stigma – usually reserved for manifestly anti-social behavior such as violent beating – to doctors and scientists operating within their professional ambit. In these circumstances, the tort or offense of Negligence is likely to offer potential plaintiffs a relatively more effective legal avenue for pursuing an alleged wrongdoing physician-investigator.

To establish a cause of action in Negligence (Malpractice) against a physician-investigator, the claimant must prove that, in the concrete circumstances of the case at bar, the defendant breached the duty of due care owing to the claimant and that the said breach resulted, with a legally sufficient causal link, in an injury suffered by the claimant. Needless to emphasize, injured patient-subjects are frequently faced with evidentiary and other constraints impeding their endeavor to convince a court of law that physician-investigators behaved negligently, namely, in violation of their obligation to act with reasonable skill according to accepted professional standards. In any event, even if ultimately successful in the judicial proceedings, suing for Assault or Negligence may only provide plaintiffs with an after-the-fact remedy, usually in the form of pecuniary compensation. Other than by way of general professional deterrence, the traditional tort and criminal law standards fail to provide an effective preventive mechanism forestalling or minimizing from the outset instances of abuse in the field of biomedical research involving human beings.

In recent years, ethics committees were constituted in various medical centers and universities in Israel. This has been done in a sporadic, piecemeal, rather improvised fashion, often to satisfy the formal requirement by funding institutions abroad that an ethics committee pass on and endorse a research protocol before a grant can be offered. The overall performance of these institutional committees cannot be assessed in depth for lack of sufficient empirical data as to their *modus operandi*. It is fair, nonetheless, to assume that, by and large, the criteria and procedures elaborated by such bodies fall short of providing a pervasive mechanism of biomedical human research control.

Towards the close of the year 1980, the Director-General of the Ministry of Health promulgated the Health Regulations[4] 1980 ([19], p. 192). This measure presents the first endeavor to devise a legislative supervisory structure in the field of human experimentation.

The term "medical experiment on humans" extends to the experimental administration of medication, radiation, or a chemical, biological, radiological, or pharmacological substance, which is bound to affect the health, body, or psyche of a person or an embryo, including the genetic code. The term also embraces the performance of any process, procedure, and investigation on a person which is not common. The Human Experimentation Regulations impose a blanket prohibition on medical experiments conducted in a hospital, unless such experiment has been authorized by the Director-General in writing and subject to the conditions set forth by him. Patently, this prohibition applies only to clinical research done in hospitals but not elsewhere. One can doubt the wisdom of restricting so severely the Regulations' scope of coverage. Why should biomedical human experimentation conducted, say, in a university department or a research institute be exempted from supervision under the Regulations? The Human Experimentation Regulations further provide that no medical experiment on humans may be conducted in a hospital in violation of the provisions of the Helsinki Declaration.[5] The Helsinki Declaration, the text of which is reproduced in the Appendix to the Regulations, has thus been incorporated into Israeli law and become the official guide to the conduct of human research in Israel. Before the Director-General may authorize a medical experiment on human beings, it must be approved by a Helsinki committee, i.e., an institutional review board at the intended hospital where the experiment is to be carried out. In addition, prior to deciding on the matter, the Director-General must obtain an opinion from the Drugs and Food Administration of the Ministry of Health or from the Supreme Helsinki Committee for Medical Experiments on Humans appointed by the Director General on a general or *ad hoc* basis. The opinion of the Supreme Helsinki Committee must be sought in the following instances: (1) an experiment concerning the human genetic code; (2) an experiment concerning the artificial fertilization of a woman; (3) any other matter as to which the Director-General wishes to be advised concerning the experiment's compliance with the provisions of the Regulations and of the Helsinki Declaration. The Committee's ten member panel includes a jurist, a clergyman, six university professors (of which at least three are physicians), the Director-General of the Ministry of Health or his representative (provided that they are licensed physicians), and the head of the Medical Association in Israel. The Director-General instruct hospitals regarding the composition of their respective Helsinki committees.[6] An application for an authorization of a biomedical experiment involving humans is submitted in writing to the Director-General by both the director of the hospital where the experiment is expected to be conducted and the physician in charge of the experiment. The application must be accompanied by a detailed research protocol, including a statement of the objectives underlying the proposed experiment as well as all tests and research endeavors already conducted within the framework of the experiment in question or of similar experiments in Israel or abroad. The Director-General may require further information regarding the intended experiment either prior or subsequent to authorizing it. He may also set conditions to his authorization and qualify or abrogate it at all times.

The Human Experimentation Regulations expressly stipulate that an authorization by the Director-General of a drug "intended for medical or other scientific research" under Regulation 29 (a)(5) of the Pharmacists Regulations [25] shall be deemed as an authorization given under the Human Experimentation Regulations. It seems that, in the wake of the promulgation of the 1980 Human Experimentation Regulations, the exercise by the Director-General of his discretion under Regulation 29 (a)(5) of the Pharmacists Regulations must now be guided by the directives set forth in the Human Experimentation Regulations. Otherwise, the Director-General could be tempted to by-pass these directives by granting *ad hoc* permits within the framework of Regulation 29 (a)(5) of the Pharmacists Regulations. Finally, it is explicitly stated in the Human Experimentation Regulations that their provisions are "in addition to any law and canons of ethics; any right, authorization or permission conferred by or under these regulations are not in derogation from any obligation imposed by any law or canons of ethics" ([19], Reg. 9). That is to say, the requirements set up by the Human Experimentation Regulations (including the prescriptions of the Helsinki Declaration) constitute the necessary minimum legal standard for the sanctioning of biomedical human research.

It is evident from the above survey that the Human Experimentation Regulations are essentially procedural in nature. Substantive standards for biomedical human research are to be found in the Helsinki Declaration, which provides a medley of principles, guiding considerations, and some procedural provisions. Like many similar national and international instruments, the Helsinki Declaration's pronouncements are nebulous in tenor and often vague in formulation. Their practical utility in guiding ordinary decision-making by those responsible for biomedical human research is rather limited. The Helsinki Declaration just fails to offer specific guidelines amenable to easy concrete application. The two principal criteria prescribed in it are the Risk-Benefit calculus[7] and the Informed Consent doctrine.[8]

The promulgation of the Human Experimentation Regulations in December 1980 has definitely heightened the concern of major medical institutions in Israel regarding the problematics of biomedical human research. Thus, for instance, the central management of Kupat Cholim (the widespread Health Service of the Israeli Labour Federation) circulated in January 1981 a lengthy document, entitled "Legal-Medical Guidelines – Medical Experiments on Human Beings" and addressed to all regional bureaus, institutes, and branches of the Health Service. The document spells out in detail the various provisions of the Human Experimentation Regulations and ventures some additional commentary. It makes cursory reference to relevant general principles of legal responsibility (i.e., Assault and Negligence under criminal and tort legal prescriptions) and to customary canons of professional ethics (notably the Helsinki Declaration, the text of which is appended to the document). The requirement of Informed Consent is accorded special attention. The document incorporates reasonably adequate model Informed Consent forms. There is also a brief allusion to the peculiar ethical dilemmas concerning experimentation on particularly vulnerable populations, such as children, old people, the chronically ill, dying patients, prisoners, and patient-subjects who are closely dependent on physician-investigators.

A year later, in January 1982, the management of the Sheeba Medical Center at Tel Hashomer issued a detailed "Procedure for the Authorization and Conduct of Research". This instrument provides, *inter alia*, for the constitution, composition and appointment procedures of a Research Committee acting, whenever human research is involved, as a Helsinki committee within the meaning of the Human Experimentation Regulations. The instrument expressly stipulates that all research done under the auspices of the Sheeba Medical Center must be guided by the Helsinki Declaration, by the guiding principles in the care and use of animals approved by the Council of the American Physiological Society, by the Human Experimentation Regulations, as well as by all relevant Israeli laws and binding canons of ethics. It further sets forth minute procedures for the approval of research proposals prescribing, *inter alia*, the content and form of research protocols (including a requirement of a risk-benefit analysis) and the structure of the proceedings conducted by the Research Committee. The text of the Human Experimentation Regulations (comprising the Helsinki Declaration) is appended to the Sheeba Medical Center instrument. It is instructive to compare this instrument with an earlier document, entitled "Guidelines to the Conduct of Research", which was adopted in March 1974 by the management of the Sheeba Medical Center. The January 1982 instrument portrays a markedly more pervasive regulatory scheme, thus attesting to the growing recognition of recent years in the need for controlling human research more closely.

The enhanced awareness of the biomedical community concerning the intricacies of experimentation in man has not passed over the university milieu in Israel. Thus, in June 1981 the Rector of Tel Aviv University appointed a Helsinki committee and assigned it the task of formulating rules and procedures regarding university-affiliated biomedical and behavioral research involving human beings. The Tel Aviv University Helsinki committee is genuinely multi-disciplinary. Among its members are a philosopher, a psychologist, and a jurist. The committee took upon itself to draft a model research application form, consisting of a detailed statement of the nature of the proposed experiment, its underlying objective and scientific significance, its logistical parameters (e.g., professional manpower, facilities, equipment, materials, and over-all research *modus operandi*), and its anticipated benefits, side-effects, and risks. Such an application form must accompany a research protocol for which the committee's approval is sought. In addition, the Helsinki Committee undertook to formulate a model Informed Consent form, expressing the voluntary participation of the subject following a disclosure of all material information – including possible risks and side-effects. The Tel Aviv University Helsinki committee also operates as an institutional review board, evaluating concrete research proposals submitted for its endorsement. In this capacity, it was asked to consider, *inter alia*, the ethical propriety of obligating psychology students to participate as subjects in psychological tests conducted by the university Psychology Department.

The Human Experimentation Regulations are definitely a step in the right direction. Considering their innovative nature, one can tolerate – for a while – the flaws in the legal arrangements they provide, and there are many such flaws. In the first place, it is submitted that such a complex and sensitive problem-area as biomedical research on human beings is worthy of treatment by the Legislature itself speaking through primary legislation.[9] The Pharmacists Regulations were promulgated by the Minister of Health and the Human Experimentation Regulations by the Director-General of the Ministry of Health. With all due respect, administrative secondary legislation is not the constitutionally proper vehicle for establishing the basic normative framework for human research.

Secondly, the present statutory scheme, which builds on these two sets of regulations, does not cover the whole field, fails to address some of the most crucial issues, and is generally lacking in consistency and depth. In particular, the control mechanism provided by it is manifestly inadequate. What we need, therefore, is an integrative and perceptive measure, grounded in primary legislation, which will come to grips with all underlying issues of the biomedical human experimentation process, such as:

(1) A possible distinction between categories of clinical trials where effective control should always be mandatory and experiments which might be exempted from close outside scrutiny (e.g., no-risk research).
(2) A possible distinction between full supervision and expedited review in certain instances (e.g., routine, minimal-risk procedures and minor variations in already approved research protocols).
(3) Special substantive and procedural prescriptions for the recruitment of particular groups of research subjects, such as notably vulnerable populations (e.g., children, prisoners, the mentally sick, and the terminally ill).
(4) An elaborate scheme of the Informed Consent process, including detailed instructions as to the essence and scope of information to be imparted to patient-subjects as well as carefully designed procedures for the effective and continuous monitoring of that process.
(5) Adequate provisions for insurance and fault or no-fault compensation of subjects injured in the course of biomedical research ([4], [6]; [2], p. 178).
(6) Appropriate standards concerning the constitution and *modus operandi* of institutional review boards and the central committee for clinical experiments (composition and appointment of members; decision-making process; manpower, budget, and other logistic facilities; documentation and records; privacy and confidentiality).
(7) Viable arrangements for continuous review and periodical reporting (particularly of modifications in research design, problems, and injuries).

As mentioned above, the Human Experimentation Regulations call for the appointment by the Director-General of the Ministry of Health of a central committee for clinical experiments, with which he should or may consult before deciding on an application to sanction a medical experiment on human beings already approved by a Helsinki committee at the hospital where it is sought to hold the experiment. The Director-General appointed such a central committee (the Central Review Board) towards the end of 1980. Its members were leading biomedical scientists and physicians from the major academic institutions in Israel as well as a clergyman and a jurist. It is convened at the discretion of the Director-General and is chaired by him. The Central Review Board is called on occasionally to deliberate and advise the Director-General in concrete instances. Given the relative scarcity of pre-formulated human experimentation standards, the Central Review Board must break new paths on virgin soil. It functions in fact as a semi-legislator, fashioning standards of a general purview in the course of scrutinizing specific research proposals. The first major case assigned to the Central Review Board was *in vitro* fertilization and embryo replacement. A qualified authorization to engage in this reproductive technique was granted by the Director-General, upon the advice of the Central Review Board, in mid-1981. Another dilemma brought to its attention was the requested administration of an unlicensed drug to a terminally ill patient.

The latter case, which reached the Israeli Supreme Court, merits brief elaboration. Mr. Reuven Ma'ayan suffered from terminal cancer. His attending physicians were unable to administer any effective treatment – Ma'ayan's illness proved unresponsive to accepted therapeutic measures. With his condition rapidly deteriorating, he sought permission to be treated with DMBG, a laetrile-related substance, unlicensed according to the Pharmacists Regulations and under development at that time by certain Israeli researchers. The Director-General of the Ministry of Health refused to authorize the use of DMBG in light, *inter alia*, of the unwillingness of the patient's attending physicians to apply the substance in question and of the similar negative position taken by the Central Review Board. The Supreme Court, petitioned in the matter by the patient and members of his family, upheld the Director-General's refusal [20].[10] A few days later, Mr. Ma'ayan died.

The Ma'ayan case attracted much public attention, complete with extensive mass media coverage and emotional debates in governmental and parliamentary circles. Many accused the Ministry of Health of formalistic insensitivity and bureaucratic delay in handling the matter.

On 11 January 1983, the Knesset (Parliament) Labour and Welfare Committee issued a statement of conclusions in the wake of the Ma'ayan affair.

Albeit endorsing the stand taken by the Director-General in refusing to approve the experimental application of the substance in question, the Committee nevertheless criticized the Ministry of Health for failing to furnish full and unobscured information to the public, such failure creating unnecessary confusion and embarassment. The Committee also expressed its concern that the proscriptions inscribed in the Pharmacists and Human Experimentation Regulations might be abused, deliberately or otherwise, thus unnecessarily hindering the progress of scientific medical research. To forestall such an undesirable eventuality, the Committee recommended to enshrine in legislation the composition and appointment procedures of the organs authorized to license drugs and approve clinical trials on humans. In the Committee's view, such legislation is bound to guarantee the scientific qualifications and full independence of such organs. The Committee also endorsed the establishment of a special Appeals Board for exceptional instances and the subjecting to scientific and public scrutiny, possibly leading to legislation, of the special predicaments entailed in the treatment of terminal patients.

The concept of Informed Consent has been repeatedly referred to above as a cardinal prerequisite to the conduct of clinical trials on humans. This oft-discussed concept is pregnant with problems, ranging from the philosophical to the pragmatic. A host of observers shares the feeling that Informed Consent in many cases is a fairy tale ([12] [13] [14] [17] [18]; [29], p. 102; [30], p. 740). It frequently means nothing beyond going through the motions, (mis?) leading the (unwary?) patient-subject to sign on the dotted line. Too many a biomedical professional genuinely believes that Informed Consent starts and ends with a signature on a piece of paper. Very little thinking, concern, and care surround the everyday process of imparting information to patients and subjects with a view to obtaining their consent to therapy or experimentation. We have hardly explored the intricacies of the doctor-patient dialogue, which is the backbone of the Informed Consent process. We are continuously shooting in the dark, indulging in unresearched assumptions as to the actual ability of patient-subjects to form a reasonably voluntary, free consent on the basis of a presumed adequate disclosure of material information. We are inclined to ignore the special Informed Consent predicaments of particularly vulnerable populations: the very young, the very old, the very poor, the very sick. We tend to belittle the significance of the factors of anxiety and dependence as inevitable emotional and psychological impediments to true Informed Consent, particularly in the context of disabling illness and excrutiating suffering. We pretend that the major obstacle to Informed Consent is the incapacity of ignorant patient-subjects to comprehend sophisticated scientific data, whereas in reality the principal stumbling block often lies in the inability of sophisticated physician-investigators to acknowledge doubt and admit ignorance. As Katz has eloquently put it, "The words necessary to communicate the certainties and uncertainties involved in participation in research can be found once one is willing to utter them" ([16], p. 785). It is submitted that, to be viable, a system of public control over biomedical experiments involving human beings must face the Informed Consent difficulties frankly, boldly, and realistically.

Lubowski Chair of Law and Biomedical Ethics
Tel Aviv University
Ramat-Aviv, Israel

NOTES

[1] Such interest is also grounded in the tenets of scientific and academic freedom.
[2] On the regulatory situation in the United States see: [1] [5] [8] [9] [10]. For comment see [28]. On professional trends in the United Kingdom see [15] [31]. Relevant German statutory directives are found in [27]. For evolving international standards see [26] [32].
[3] Formerly in exercising his discretion under this regulation, the Director-General of the Ministry of Health was advised by a Clinical Trials Committee established by him for this purpose. The Committee comprised, in addition to the Director-General himself, a clinical pharmacologist, two medical clinicians engaged in research, and several pharmacists from the Ministry of Health who deal, *inter alia*, with drug licensing. For a general survey of the *modus operandi* of the Clinical Trials Committee see [23].
[4] See Appendix in this volume.
[5] The Declaration of Helsinki – Recommendations Guiding Medical Doctors in Biomedical Research Involving Human Subjects – was adopted by the 18th World Medical Assembly at Helsinki, Finland, in 1964, and was revised by the 29th World Medical Assembly in Tokyo, Japan, in 1975. The Knesset (Israeli Parliament) Public Services Committee endorsed the Helsinki Declaration in a resolution on medical experiments with humans that passed in January 1967.
[6] In a circular issued to all hospital directors in Israel in March 1981, the Director-General instructed that a Helsinki committee should be comprised of five expert physicians appointed by the directors of the hospital in question, including at least three heads of

hospital departments representing different medical disciplines, one representative of the hospital's medical management, and at least one expert internist. No provision was made for mandatory lay participation (of, e.g., an ethicist, clergyman, jurist, or social worker) in hospital Helsinki committees.

[7] See Appendix, in this volume, II *Medical Research Combined with Professional Care (Clinical Research)* (1) (2).

[8] See Appendix, in this volume, I *Basic Principles* (9).

[9] For an example of such a legislative endeavor see [21]. The draft private bill is based on a paper written by Mr. Y. Natanson, a law student, in a seminar on Law and Medicine conducted by the present writer at the Faculty of Law, Tel Aviv University. For an earlier attempt at formulating a regulatory human experimentation statutory measure see [22].

[10] For a critical review of the Court's decision and reasoning see ([24], p. 445).

BIBLIOGRAPHY

1. The Belmont Report, *Ethical Principles and Guidelines for the Protection of Human Subjects of Research*: 1978, DHEW Publication No. (OS) 78–0012.
2. Calabresi, G.: 1969, 'Reflections on Medical Experimentation in Humans', in P. A. Freund (ed.), *Experimentation with Human Subjects*, Russell Sage Foundation, New York.
3. Carmi, A.: 1977, *The Doctor, the Patient and the Law*, 25–56, 61–122 (Tel Aviv, in Hebrew).
4. Casebeer, K. M.: 1976, *Injured Research Subjects: What Do We Owe Them?*, 2, *Quarterly Report of the Kennedy Institute for the Study of Human Reproduction and Bioethics* 5–9.
5. Code of Federal Regulations, Title 45, Public Welfare, Department of Health and Human Services, National Institutes of Health, Office for Protection from Research Risks, Part 46 – *Protection of Human Subjects*, Revised as of March 8, 1983.
6. Note: 1975, 'Comparative Approaches to Liability for Medical Maloccurrences', *Yale Law Journal* **84**, 1141.
7. Deutsch, E.: 1979, 'Medical Malpractice: Informed Consent and Human Experimentation in Western Europe', *International Journal of Medicine and Law* **1**, (1), 81.
8. *Federal Register*, Vol. 46, No. 17, January 27, 1981, Rules and Regulations, pp. 8942–8980, Department of Health and Human Services, Food and Drug Administration, *Protection of Human Subjects & Informed Consent*.
9. *Federal Register*, Part II March 22, 1982, Department of Health and Human Services, Office of the Secretary, *Exemption of Certain Research and Demonstration Projects from Regulations for Protection of Human Research Subjects*.
10. *Federal Register*, Part IV, Department of Health and Human Services, Public Health Service, July 28, 1983, *Protection of Human Subjects, Reports of the President's Commission for the Study of Ethical Problems in Medicine and Biomedical and Behavioral Research, Notice of Availability and Request for Public Comment*.
11. *Final Report of the Tuskegee Syphilis Study Ad Hoc Advisory Panel*, U. S. Department of Health, Education and Welfare, April 1973, Code of Federal Regulations, Title 45, Public Welfare, Department of Health and Human Services, Part 46 – *Protection of Human Subjects, Revised as of January 26, 1981*.

12. Fletcher, J.: 1969, *Human Experimentation: Ethics in the Consent Situation*, in C. C. Havighurst (ed.), *Medical Progress and the Law*, Oceana, New York.
13. Goldstein, J.: 1975, 'For Harold Lasswell: Some Reflections on Human Dignity, Entrapment, Informed Consent, and the Plea Bargain', *Yale Law Journal* **84**, 690–698.
14. Halligan, P. D.: 1980, 'The Standard of Disclosure by Physicians to Patients: Competing Models of Informed Consent', *Louisiana Law Review* **41**, 9.
15. *Health Service Circular* (Interim Series) 153, June 1975, sent by the Department of Health and Social Security to Regional Authorities and others, and incorporating the 1973 *Report* of a committee of the Royal College of Physicians on the Supervision of the Ethics of Clinical Research Investigations in Institutions.
16. Katz, J.: 1973, 'The Regulation of Human Research – Reflections and Proposals', *Clinical Research* **21**(4), 785–791.
17. Katz, J.: 1977, 'Informed Consent – A Fairy Tale?: Law's Vision', *University of Pittsburgh Law Review* **39**(2), 137–174.
18. Katz, J. and Capron, A.: 1975, *Catastrophic Diseases: Who Decides What?*, Russell Sage Foundation, New York.
19. Public Health Regulations (Human Experimentation), December 1, 1980, 4189 Kovetz Hatakanot, (Secondary Legislation).
20. Ma'ayan, R. *et al.* v. the Director-General of the Ministry of Health: 1982, **36** (II) *Piskei Din* (Supreme Court Judgments).
21. Medical Research (Experiments with Human Beings) proposed law, by member of the Knesset Professor A. Rubinstein (November 1980).
22. Medical Research Law, drafted 1974 by Judge A. Carmi and Professor A. Atzmon under the auspices of the Society for Medicine and Law in Israel.
23. Nir, I.: 1983, 'Clinical Trials in Israel – Criteria for Approval', *Israel Journal of Medical Sciences* **19**(5), 445–448.
24. Parush, A.: 1982, (Hebrew) 'The Ma'ayan Case – Legal & Moral Aspects', **12** *Mishpatim* (the Hebrew University Law Review), 530.
25. Pharmaceuticals, May 23, 1986: 4933 Kovetz Hatakanot (Secondary Legislation).
26. Proposed International Guidelines for Biomedical Research Involving Human Subjects, World Health Organization and the Council for International Organizations of Medical Sciences, CIOMS Geneva, 1982.
27. Reform of Drug Legislation of the Federal Republic of Germany of 24 August 1976 (Federal Law Gazette I, p. 2445), Sixth Section s. 40–42 ("Protection of the Human Being During the Clinical Trial").
28. 'Regulations Governing Research on Human Subjects – Academic Freedom and the Institutional Review Board', Dec. 1981, *Academe* **67**.
29. Robertson, G.: 1981, 'Informed Consent to Medical Treatment', *Law Quarterly Review* **97**, 102.
30. Somerville, M.: 1981, 'Structuring the Issues in Informed Consent', *McGill Law Journal* **26**, 740.
31. Statement by the Medical Research Council: 1964, 'Responsibility in Investigations on Human Subjects', *British Medical Journal* **2** (July 18), 178–180.
32. United Nations, Economic and Social Council, Commission on Human Rights, Thirty-first session, E/CN. 4/1173 (18 February 1975) 9–10, 15–21.

SECTION III

ETHICAL AND EPISTEMOLOGICAL ISSUES IN RANDOMIZED CLINICAL TRIALS

H. TRISTRAM ENGELHARDT, JR.

DIAGNOSING WELL AND TREATING PRUDENTLY: RANDOMIZED CLINICAL TRIALS AND THE PROBLEM OF KNOWING TRULY

I. INTRODUCTION

All human knowledge is tentative and provisional, and medical knowledge is no exception. Explanations of diseases and disease processes are tied to patho-anatomical, physiological, and microbiological theories that change with time. So too, the best of standard care is open to revision as better knowledge comes to hand. Medicine is, as all other human endeavors, bound to the history and culture of its setting. Moreover, since the nature of the world is in part opaque to, or at least recalcitrant before the inquiries of human reason, explanations of medical problems and attempts to remedy them are always less than ideal. In fact, the limitations of human knowledge destine medical interventions to being not only less than fully efficacious, but also to bringing risks of morbidity and mortality that might better have been avoided.

We can now look with amusement at our ancestors who were bled, purged, and blistered for no benefit and often to their positive harm. Faced with the debilities, pains, and vexations of illness, and confronting the fear of death, people importunately, if not desperately, seek cures. Medicine as the institutionalization of these attempts was understandably impelled to pursue forms of treatment that we now in retrospect know to have been misguided. It is difficult for humans to face disease, deformity, and premature death without attempting to do something. Moral and intellectual virtues of restraint are required if one is not to accept a treatment as efficacious if it has frequently been associated with the cure of dangerous or painful diseases, even if one has not established that this association is more than coincidental. It takes similar intellectual virtue to be skeptical of the forms of treatment that are accepted by one's teachers and one's professional community. The current controversies regarding the comparative efficacy of conservative vs. surgical treatment of coronary artery disease (See, e.g., [12], [41], [53]), or the past difficulty in rejecting bleeding as an established treatment [48] demonstrates the need for an intellectual discipline of looking critically at what appears at first blush to be successful, or what has been accepted as standard treatment.

There are, however, many obstacles to developing this intellectual discipline.[1] First and foremost, there is the problem of random occurrences. Some individuals will be about to be cured from other causes, just as one applies a totally inefficacious treatment.[2] As a result, one will be tempted fallaciously to interpret a patient's restoration to health as a result of therapeutic interventions. Moreover, physicians appear to remember their therapeutic triumphs more clearly than their failures ([58], [19]). A failure to understand the natural history of diseases may also confuse a cure with the natural termination of a self-limited disease process [7]. Beyond that, medical intervention of whatever sort is often tied to a powerful placebo effect. People often feel better even when the treatment has nothing specifically to do with their improvement in well-being [63]. Finally, the psychology of discovery is such that people tend to see the world in part at least as they expect it to be.[3] Even the most intellectually steadfast of clinicians is subjected to his or her own clinical biases. The results of these biases can be misperceptions of data and of the efficaciousness of clinical interventions. One might think here of recent attempts to decide whether radical or simple mastectomies for carcinoma of the breast offer women a greater chance of being cured from the disease ([14], [51], [61], [62]). Or one might think of 19th century examples such as the disease of masturbation. Thoughtful individuals such as Sigmund Freud, Jonathan Hutchinson, and C. E. Brown-Sequard concluded that masturbation was the cause of serious mental and physical sequelae [20]. In fact, individuals such as Brown-Sequard recommended invasive treatment, including cauterization of the clitoris to preserve the health of patients.[4]

It is very difficult to know truly in medicine; and since medical knowledge is knowledge in the service of action, false knowledge can lead to well-intentioned interventions causing patients needless harm. To summarize, random variation, imperfect knowledge of the history of diseases and their often self-limiting or episodic character, as well as the placebo effect may lead to false *post hoc ergo propter hoc* conclusions with regard to the efficaciousness of medical treatments. Selective memory and distortions in the psychology of discovery compound these difficulties. In that it is unlikely that individuals will refrain from using those treatments they believe may be helpful, one is pressed to develop means of rigorously and critically assessing claims to medical knowledge, especially with regard to the efficaciousness of particular therapeutic modalities. Experimentation with human subjects is thus tied to a project of systematically assessing medical claims so as to choose those therapeutic approaches that will do the least harm and the most good.

I will attempt here to explore the relationship between critical understandings of medical knowledge and the development of systematic practices for testing medical claims and possibilities for medical intervention. My goal will be to suggest what ought to be an obvious thesis, that the practice of experimentation on human subjects is *prima facie* good. All else being equal, it is good and appropriate to encourage research with human subjects where such can lead to a better understanding of what will serve as efficacious treatment (including preventive medicine). To avoid false conclusions, we will need to fashion such research around such devices as randomized clinical trials, often including double blind studies. The practice of human experimentation pursues a set of important goals that are recommended by the very character of medical treatment. The use of medicine involves goals and goods (e.g., avoiding dangerous treatments), which recommend the practice of research involving human subjects. Communities that employ scientific methods should encourage their members to participate in well-structured human experimentation, even if they may not compel such participation.[5]

II. AN ONUS AGAINST EXPERIMENTATION

If one looks to history, one often finds what appears to be an onus against the physician who departed from standard modes of medical treatment. As Darrel Amundsen shows in a study of the regulation of physicians in the late Middle Ages, the Christian Church looked with suspicion on increased experimentation by physicians and surgeons during the late 15th and early 16th centuries ([1], [2]). For example, Batholomaeus Fumus, in his *Summa Armilla* of 1538, holds that physicians sin "if they supply a doubtful medicine for a certain one, or do not practice in accord with the art, but desire to practice following their own stupid fancy, or make experiments and such like, by which the patient is exposed to grave danger" ([1], p. 35). Other criticisms, such as those of Batholomaeus Chaimis, in his *Interrogatorium sive confessionale* of 1474, and Cajetan, in his *Summula peccatorum* of 1525, condemned experimentation as a mortal sin if done deceitfully or on the poor ([1], p.

35). In this second case, the concern is not so much with experimentation *per se*, as with experimentation that appeared exploitative. Unquestionably, the suspicion that the poor and defenseless would be exploited has, sadly, been well-founded. One finds, for example, in Katz's mammoth review of issues in human experimentation, vignettes of 19th and 20th century cases of abuse [46]. One need but think of recent examples such as the Tuskegee syphilis study to realize that even outside of Nazi Germany, problems in the exploitation of patients have reached into the midst of our century.[6]

However, in addition to well-founded caution regarding the possibilities for abuse, there has remained a suspicion with regard to experimentation itself. As William Curran has shown ([16], p. 403), from cases such as *Carpenter v. Blake* [13] in 1871, to *Brown v. Hughes* [11] in 1934, some American courts regarded experimentation as equivalent to a rash action on the part of a physician, though, as Curran also argues, this probably is not a fully representative view of the law during this period. Still popular books concerning physicians and the law ([57], p. 381) advised physicians that they would be engaging in experimentation at the risk of civil liability. In fact, a contemporary defense of this attitude, from a moral point of view, is to be found in a 1969 *Daedalus* article by Hans Jonas [43], revised and reprinted in 1970 [44], in which he describes human experimentation as *prima facie* dehumanizing, even when the subject consents. Since human experimentation requires using a patient as a means, albeit given consent not as a means merely, Jonas finds this to be morally suspect. He argues that the "compensations of personhood are denied to the subject of experimentation, who is acted upon for an extraneous end without being engaged in a real relation where he would be the counterpoint to the other or to circumstance. Mere 'consent' (mostly amounting to no more than permission) does not right this reification. Only genuine authenticity of volunteering can possibly redeem the condition of 'thinghood' to which the subject submits" ([44], pp. 3–4). As Jonas puts it in his *Daedalus* article, "What is wrong with making a person an experimental subject is . . . that we make him a thing – a passive thing, merely to be acted upon . . ." ([43], p. 235). Jonas, who places mere volunteering on a moral par with conscription ([44], pp. 16–17), sets the criteria for "authentic volunteering" extremely high.[7] Moreover, he sees experimentation as a means to a merely optional good, the conquest of disease and the postponement of death ([44], pp. 28–29). He shows little appreciation of the dilemma of physicians, who today are in many areas in the position of their colleagues in the past who bled patients in therapeutic regimens that caused much more harm than would have been necessary, had they had a better knowledge of disease and of the efficaciousness of alternative forms of treatment. Therefore he fails to speak to the strategic difficulties of physicians. Today and in the future, physicians will likely always be enlisted by many to intervene maximally to cure serious, painful, or life-threatening diseases. However, with inadequate knowledge of the nature ow pathological processes or of optional treatment, they will be enlisted to cause more harm than would be necessary, given better knowledge.[8]

The search for better knowledge through human experimentation is not simply the attempt to conquer ever more diseases and postpone death ever further. It is part also of a protection against well-established forms of treatment, as well as new forms of treatment, which bring with them high risks of morbidity, if not mortality, and which with a better knowledge of things would be rejected, as bleeding was in the 19th century. Human experimentation must be understood within the context of a human compulsion to treat diseases, often with heroic and risky regimens, even in the presence of inadequate knowledge. With the commitment to risk one's well-being by supporting an institution that develops painful and life-threatening interventions, one may have as well committed oneself to the reasonableness of exposure to risks as a part of research in order to gain knowledge that will, over the long run, lower one's risks in such an institution. From the point of view of patients and prospective patients, participating in human experimentation, under such circumstances, may be a reasonable part of a strategy to maximize one's benefits in encounters with medicine. It will be a form of a premium paid to insure oneself against some risks in an endeavor fraught with risks. It becomes difficult, if not impossible, to establish an onus against human experimentation once one has committed oneself to the aggressive use of medicine in the control of pain, disability, and the postponement of death.

III. TESTING AND EXPERIMENTING

In the same century that Cajetan wrote with suspicion of human experimentation, Francis Bacon (1561–1626) was born. Bacon, who drew a distinction between men of experiment and men of dogma, enjoined the employment of experimentation to force from nature her secrets [3]. Bacon's interest in experimentation and natural history was to have a direct influence on physicians into the late 18th century. First and foremost was Bacon's influence on Thomas Sydenham (1624–1689). Sydenham drew from Bacon both a commitment to a disciplined observation of nature in order to develop reliable natural histories of diseases, as well as a belief that through experimentation one would be able to establish the long-sought *methodus medendi*, a mode of treating individuals that would be well-founded and successful. As a Baconian, he attempted to fashion natural histories of disease after having first swept away what he took to be the distorting force of medical theory [60]. He sought as well, taking a term from Bacon, to develop *experimenta fructiferra*. For example, he enjoined experimentation with medications, not in order simply to establish yet one more remedy for the pharmacopaeia, but rather in order to develop more comprehensive and simplified modes of therapy.

One thus finds Sydenham offering a dual injunction regarding the improvement of medicine. On the one hand, he recommends the careful description of the natural histories of diseases. On the other hand, he urges the development of a "fixed and every way complete method of cure . . . sufficiently established and verified by a complete number of experiments, and found effectual to cure any particular disease" ([59], p. xii). Sydenham argued that it was important, but not sufficient, for a physician "to attend carefully to the particular effects both of the method and medicine he uses in curing diseases, and to set them down for the ease of his memory, as well as the improvement of his knowledge; so that at length, after many years experience, he may fix upon such a method of curing any particular disease, as he need not in the least depart from" ([59], p. xii). Sydenham found, however, that such endeavors, at least as he had become acquainted with them, were not "so useful . . . for if the observer only intends to inform us that a particular disease hath yielded once, or oftener, to such a medicine, of what advantage is it to me that a single medicine which I knew not before, is added to the immense stock of eminent medicines, that we have long been pester'd with?" ([59], pp. xii–xiii). To understand this complaint, one must appreciate the chaotic nature of the *materia medica* of the time. Sydenham is petitioning for an empirically well-founded way to render treatment not only more efficacious, but simpler as well. He thus asks whether it would not be better that he "lay aside all other medicines [and] . . . use only this . . ." so that the virtues of a putatively efficacious new medicine could be "approved by numerous experiments" ([59], p. xiii).

Sydenham's actual injunctions about particular remedies show the difficulties in gaining a reliable and systematic basis for therapy. Along with his usually careful injunctions regarding the use of the Baconian method, one finds less careful recommendations. One might mention, for example, Sydenham's endorsement of Dr. Goddard's drops prepared by Dr. Goodall ([59], p. xx).

These difficulties with therapeutics remained into the 19th and early 20th centuries. Consider some of the recommendations by William Cullen (1710–1790), who was influenced both by Bacon and by Sydenham. In *First Lines of the Practice of Physic*, which was reprinted and used widely in early 19th century in America, Cullen recommended this treatment for hepatitis: "The cure of this disease must proceed upon the general plan; by bleeding, more or less, according to the urgency and pyrexia; by the application of blisters; by fomentations of the external parts in the usual manner, and of the internal parts by frequent emollient clysters; by frequently opening the belly, by means of gentle laxatives, and by diluent and refrigerant remedies" ([15], p. 163).

Medicine has, since the death of Cullen, been changed by the development of what we have come to term the basic sciences. Indeed, even before the 19th century one had the *Sepulchretum* of Theopile Bonet (1620–1689) [9] and the *De Sedibus et causis morborum per anatomen indagatis* of Morgagni (1682–1771) [50]. However, the 19th century, as Foucault has shown [34], was a turning point in the understanding of medicine. Medicine turned inward to the study of the body through the work of Xavier Bichat (1771–1802) [5] and those who followed him, who offered a way of bringing unity to the bewildering mass of clinical data [22]. With these changes there was a change in the understanding of what counted as good empirical medical science. Physicians and physiologists began to depart from the simple observations of the naturalists and to become self-conscious manipulators of causal factors. It became clear that to learn from nature, one would need not only as an intellectual exercise to free oneself from theoretical prejudices, as Sydenham had insisted following Bacon, but also to develop modes of standardizing experiments and carefully isolating the role of particular causal variables on outcomes.[9] A new sense of experimentation developed. It was not simply that the gaze turned inward to study corpses or bodies in order to understand clinical findings, but the sense of experimentation changed as well with a growing appreciation of the need to be disciplined in observation. Here one might think of the assessment by P. C. A. Louis of the efficaciousness of bloodletting as a therapy [48], or of J. Gavaret's introduction of the notion of confidence limits to medical observation [35]. As Henrik Wulff has shown, it is during this period that an attempt to establish scientific method in medical research developed as we know it.[10] In the American experience one might note the study by Pliny Earle, *The Curability of Insanity*, in which he analyzed and showed to be false previous claims about what hospital reports had maintained regarding the curability of insanity [18]. This was the period as well in which interest in logics of medicine developed, as seen in attempts to assess the reliability of reasoning in medicine ([6], [8], [56]).

During this period of time, therapeutic trials were undertaken such as those of Edward Jenner's 1796 inoculation of the 8-year-old James Phipps with cowpox ([42], pp. 32–35), and Louis Pasteur's 1885 treatment of the 9-year-old Joseph Meister with an anti-rabid vaccination [17]. One might note here as well Walter Reed's work with the Yellow Fever Commission, which in 1900 demonstrated that the *aedes aegypti* (then called *culex fasciatus*) was the vector for yellow fever. Reed, in these investigations, developed a consent form, which compares well not only with much experimentation in the 19th century[11] (which was often done with little regard to the rights of the subjects), but with current procedures as well [4].

The very notion of medicine as a rational endeavor was coming to include the controlled manipulation of causal factors in humans in order to understand the nature of disease and its treatments. One is offered the culmination of a major conceptual evolution. Medicine had moved from a tolerance of dogma, to a Baconian emphasis on careful observation, to the controlled manipulation of causal factors, and finally to the use of statistical methods. Medicine had come to augment clinical pathological correlations, which relied on the experiments of nature, with controlled experiments on living human subjects. Through these steps the very notion of rationality in medicine changed. It is only against the background of this development that one can appreciate current concerns of physicians to find ways of avoiding over- or undertreatment, false positive and false negative diagnoses. As one appreciates choices of diagnosis and treatment as attempts to select interventions that minimize harms and maximize benefits, one is confronted with the importance of controlled statistical appreciations of the bases for intervention. Medicine has become explicitly an educated attempt to play the odds for the benefit of patients. However, to do well in such a practice, one must know the odds well. Disciplined statistical studies become unavoidably part of such an endeavor. This has often included a clinical pragmatism or instrumentalist understanding of the physician's predicament as one of construing the data so as to authorize interventions in ways that will do the least harm and the most good ([36], [52]). The very ways in which medical data are given their meaning through being interpreted as signs of outcomes or as indications for treatment are dependent on a set of judgments regarding causal relationships and well-established past associations of events. Those judgments regarding causal associations depend on past observations, both disciplined and undisciplined. The more trustworthy and comprehensive those observations, the more useful and trustworthy the therapeutic choices will be, which are based upon them.

IV. TREATING WELL THROUGH KNOWING WELL

The issue of human experimentation arises not (1) in an institution dedicated only to pure knowledge, or knowledge for knowledge's sake, though such does exist outside the bounds of clinical medicine, or (2) in an institution dedicated to intervening only some time in the future to treat diseases and postpone death. Rather, it arises in a social institution already committed to such interventions and which as a consequence seeks knowledge not for knowledge's sake, but for its immediate usefulness. The question then is how and under what circumstances one may manipulate those interventions in order to decrease risks. To fashion a simile, it is like asking about the morality of employing different maneuvers to decrease injuries and deaths while engaged in mountain climbing. One might imagine individuals who have already begun a climb, and who have therefore in a real sense decided that the climb and its risks are worthwhile. If on the way a serious question arises regarding the reliability and trustworthiness of the equipment, one will need to decide on means to test it. If after assessing the modes for testing that are available, the climbers decide that the most reliable approach is to distribute different sorts of equipment at random to members of the climb and then check after some time has gone by and compare experiences, one comes close to capturing the role of experimentation in medicine.

It is clear that ethical issues are at stake. First, insofar as respect of the freedom of persons is a minimal condition for the notion of morality ([24], [25], [27], it follows that one may not conscript unwilling subjects. The mountain climbers will usually not be able to find willing non-mountain climbers to participate, just as one is unlikely to find a great number of Christian Scientists to volunteer for medical experimentation. To prevent the practice of human experimentation from becoming a means for imposing the wishes of some on others by force, and therefore becoming an immoral practice, one requires the free consent of those who participate. To be free, such consent must be as informed as the subjects wish to have it and the experimenters are able to make it. One might think here of the principle of respect for persons which the National Commission for the Protection Of Human Subjects of Biomedical and Behavioral Research in its *Belmont Report* underscores as the first of its three basic ethical principles [54]. A survey of past major codes on human experimentation reveals this to be one of the prominent and key moral principles governing experimentation [21].

Experimentation with humans should, as most codes of human experimentation emphasize, offer benefit to the particular subjects or to society [21]. Here one would wish to signal the second basic ethical principle of the National Commission for the Protection of Human Subjects, namely, that of beneficence ([54], pp. 6–8). What will count as benefits and harms must be seen against the backdrop of individuals already involved in a practice that carries with it established exposures to risks in the service of achieving the goals of improved health and longer life. The establishment of a practice in medicine of recruiting volunteers to participate in human experimentation, as Hans Jonas suggests [43], would not be alien to the purposes and goals of the individuals recruited. Instead, the development of such practices should in fact be a fulfillment of the goals of those who seek medical care. This argument should succeed as well, in fact in particular, for randomized clinical trials, including double-blind clinical trials, insofar as these provide the best information about the benefits and costs of different ways of characterizing the problems of patients, and acting to treat them.

Arguments in favor of randomized clinical trials and double-blind clinical studies develop out of an appreciation of the difficulties of judging objectively regarding the truth of knowledge claims in medicine. They appear alien only insofar as the intellectual disciplines recommended to us by R. A. Fisher ([31], [32]) and others ([39], [40]) regarding the statistical requirements for valid observations force us to reconceive the project of gaining knowledge. For the most part, individuals speak of knowledge in everyday life, as well as in medicine, in ways that belie the strategic difficulties involved in framing objective characterizations of nature. They tend to downplay the probabilistic character of empirical knowledge claims. Regarding issues involving life and death, individuals, at least in our culture, often feel ill at ease in confessing uncertainty with respect to whether their choices of treatment are well-founded and rationally justified. In addition to the uneasiness evoked by uncertainty, it may well be the case that the placebo effect of the physician is enhanced by a feigned certainty. The ability to assure those in distress is surely undermined by uncertainty. In any event, individuals generally do not conceive of medicine as an attempt to play the odds in favor of the patient under circumstances where it is often unclear what the odds are.

However, when it is not clear which of a number of alternative treatments is the best, it would seem difficult to argue that it would not be in the interest of patients in need of such treatment, or possibly in need of such treatment, to endorse a systematic manner of determining the best treatment. Randomized clinical trials and double-blind studies can thus be a reasonable strategy for the pursuit of goods endorsed or endorsable by at least a certain subset of individuals. One should also note here that free and informed consent is not a bar to such endeavors. One can consent to not knowing in advance what treatment one will receive and to having treatment assigned randomly. Just as in poker, knowing that others will attempt to deceive as part of the game defeats one's claim that one has been lied to, so too if subjects have consented in the absence of coercion, and with an explanation of the general character of the uncertainties and deceptions involved, the autonomy of the subjects has not been violated. If the antecedent moral requirement of gaining the consent of subjects is fulfilled, one has entered a sphere of moral considerations where one is contemplating acts that will at least not conflict with the minimal notion of ethics as the fabric of a peaceable community.

Here one must recognize a major problem in ethics, namely, of being able to forward a convincing argument for a particular view of the good life. Insofar as one despairs of such general arguments succeeding, one will be pressed to acknowledge the core of ethics as concerned with procedures that protect the rights of moral agents to self-determination. Even if one cannot discover what the concrete goods of persons are, one can at least recognize that the very notion of ethics as a means of resolving conflicts without direct and basic appeal to force requires mutual respect of those involved: hence, the cardinal role of free and informed consent not only in human experimentation but in medical practice generally. When individuals consent to joint action involving only themselves directly, not only is there a presumption in favor of the licitness of such action, but the individuals can indeed justify their actions in terms of the most generally defensible element of ethics: mutual respect.[12]

Beyond the issue of mutual respect, one can identify the general structure of a moral practice involving, *inter alia*, benefits pursued non-egoistically and altruistically. The practice also involves non-moral goods such as the absence of pain and deformity, full range of expected function [23], and assurance against premature death. By regarding issues of beneficence in human experimentation we can distinguish among three genres of experimentation: (1) experimentation done for the possible direct benefit of the particular patient-subject; (2) experiments likely to be of direct benefit to a class of individuals with which the subjects identify themselves (i.e., "identify" in the sense that the subjects seek the good of a particular class; when such identification is with the community of humans or persons generally, it reaches a fully altruistic level); and (3) human experimentation recognized by the subjects to bear no fruit of direct benefit to them. Within the last genre I have in mind experiments that the subjects conceive to be for knowledge's sake and not likely to be geared into institutions that would lead to benefit for the subjects, or for a class of individuals with which the subjects identify. In such cases the grounds for participating may spring from a commitment to knowledge or from a recognition that research in the end has profound implications for clinical medicine. The grounds for participation here, as with other genre, may not all be moral grounds. One might think of individuals who volunteer for experiments because of pay, escape from boredom, or, in the case of individuals volunteering for experiments in human sexuality, sexual release. The presence of free and informed consent should be sufficient to legitimize all such endeavors. However, as a matter of public policy one may wish to support or encourage only that research that in general supports the goods of the community.

The considerations I have offered support the argument that a large number of the experiments with human subjects, including those involving randomized clinical trials, fall within the first two genres. They support either the goods of the particular individuals involved, or at least the goods of the class of individuals with which the subjects identify. Insofar as individuals seek from medicine ways of curing and preventing diseases and disability and postponing death, such research would be far from alien. In fact, one can say of individuals in such circumstances that they *ought*, ceteris paribus, to participate in such experimentation, including randomized clinical trials, for such participation supports the goods they have endorsed. "Ought" here is a conditional "ought", which signals that a practice of consenting to participate in research of X sort would support goods endorsed by the individuals involved, which goods are in addition moral goods in the sense of existing within a matrix of altruism of sympathy towards others. If, for example, one's general account of ethics is utilitarian, then one would appreciate this oughtness as that associated with a practice that if properly employed will redound to the benefit of the greatest number. The moral force here is teleological and expresses the relationship between a practice and a set of goods, which includes moral goods. Whether the "ought" expresses obligation or endorsement of a supererogatory good depends on the context.

V. SUMMARY

An examination of the practice of human experimentation, even that experimentation involving randomized clinical trials and double-blind studies, shows that it is not an endeavor alien to the general moral commitments of medicine or those who come to medicine. Or at least a great proportion of such experimentation can be seen as integral to the endeavors of medicine. It is part of gaining sufficient knowledge to diagnose well and understand what it is to choose a mode of treatment prudently. As such, human experimentation is instrumental to the goods and goals of medicine. Insofar as those are affirmed, one affirms such experimentation as well. Again, this argument will not encompass all human experimentation. But it will identify a great proportion of those attempts to decide among competing alternative treatments, as well as among alternative ways of characterizing clinical problems.

I have forwarded this sketch to close the distance between human experimentation and the general endeavors of medicine. Characterizations such as Hans Jonas's, which see human experimentation as illicitly reifying the subject, must either be modified or seen as criticisms of medicine generally. They fail to appreciate the tentative and provisional nature of medical knowledge and therefore why strategies such as human experimentation, including randomized clinical trials, are embedded in the general enterprise of medicine. Indeed, I have attempted to embed the practice of experimentation in a general account of rational behavior under the circumstance of incomplete and ambiguous information, and given an interest in intervening to prevent and treat diseases. Though my account does not demonstrate that there is an enforceable obligation to participate in human experimentation (my arguments regarding autonomy show that such putative obligations could be defeated [27]), it does indicate that there is a sense in which one *ought* to participate. That "ought" is one embedded in a teleology that frames the general commitments of beneficence in medicine. Insofar as one affirms the end, it is at least incumbent on one to account for not having affirmed the means.

The Baylor College of Medicine
Houston, Texas, U.S.A.

NOTES

[1] A very helpful survey of these issues is provided by Henrik Wulff in his *Rational Diagnosis and Treatment* [64], especially chapters 9 and 10.

[2] The problem of random variation was in the past poorly appreciated. Only in the 19th century did physicians begin systematically to consider the likelihood that their sample of outcomes might not represent what would be the long-range experience. See, for example, [29], [30]; [40], pp. 6–13.

[3] Both in the philosophy of science and the philosophy of medicine there has been an increasing appreciation of the fact that the expectations of investigators mold in part the ways in which they appreciate "facts". See, for instance, Ludwik Fleck [33]; also see, [37], [38].

[4] There were, in fact, surgical treatments recommended, including removal of the clitoris. See, for example, Baker Brown's *On the Curability of Certain Forms of Insanity, Epilepsy, Catalepsy, and Hysteria in Females* [10].

[5] Richard McCormick, for example, in defending the use of children in human experimentation, developed an argument that individuals have a general moral obligation to society to participate in human experimentation. He sees this as the basis for allowing parents to "volunteer" their children who are below the age of consent [49]. He employs the wrong argument in order to secure the right conclusion [27].

[6] According to J. H. Jones, the U.S. Public Health Services Division of Venereal Diseases published a series of 13 articles from 1936 to 1973 in various journals reporting their findings from the Tuskegee Syphilis Study. For a list of these publications, see [45], pp. 257–258.

[7] For voluntary consent to be fully authentic, Jonas appears to require that the subjects have the experimenter's interest in the endeavor, and, as far as possible, the experimenter's level of knowledge of the issues at stake in the experiment. It is for this reason, among others, that Jonas regards double-blind studies involving patients as a betrayal of trust ([43], p. 240). I presume here that he is cognizant of the fact that individuals in such studies can be informed that there will be random assignments, including a placebo group.

[8] One might think of the difficulties involved, for example, in demonstrating even to reasonable individuals that laetrile is not an efficacious drug [28].

[9] For a very thoughtful study of this change in the understanding of experimentation in neurology and neurophysiology, see Robert Young's *Mind-Brain Adaptation* [65].

[10] Wulff shows the importance of Gavaret in the development of a new disciplined sense of observation ([64], pp. 131–134).

[11] The following is the consent form employed by Reed.

The undersigned,   being more than twenty-five years of age, native of Cerceda, in the province of Corima, the son of   here states by these presents, being in the enjoyment and exercise of his own very free will, that he consents to submit himself to experiments for the purpose of determining the methods of transmission of yellow fever, made upon his person by the Commission appointed for this purpose by the Secretary of War of the United States, and that he gives his consent to undergo the said experiments for the reasons and under the conditions below stated.

The undersigned understands perfectly well that in case of the development of yellow fever in him, that he endangers his life to a certain extent but it being entirely impossible for him to avoid the infection during his stay in this island, he prefers to take the chance of contracting it intentionally in the belief that he will receive from the said Commission the greatest care and the most skillful medical service.

It is understood that at the completion of these experiments, within two months from this date, the undersigned will receive the sum of $100 in American gold and that in case of his contracting yellow fever at any time during his residence in this camp, he will receive in addition to that sum a further sum of $100 in American gold, upon his recovery and that in case of his death because of this disease, the Commission will transmit the said sum (two hundred American dollars) to the person whom the undersigned shall designate at his convenience.

The undersigned binds himself not to leave the bounds of this camp during the period of the experiments and will forfeit all right to the benefits named in this contract if he breaks this agreement.

And to bind himself he signs this paper in duplicate, in the Experimental Camp, near Quemados, Cuba, on the 26th day of November nineteen hundred.

On the part of the Commission: The contracting party,

Walter Reed
Maj. & Surg., U.S.A.

This consent form is with permission from the Philip S. Hench – Walter Reed – Yellow Fever Collection, Historical Collections, The Claude Moore Health Sciences Library, University of Virginia Medical Center. I wish to thank Dr. William Bean for calling it to my attention.

[12] I am advancing here an argument that I have developed at length elsewhere, that respect of freedom functions as a condition for the possibility of the general grammar of ethics. I am in debt to the arguments of Robert Nozick for his distinction between freedom as a value and freedom as a side constraint [55]. However, unlike Nozick, my argument has a transcendental character in forwarding freedom as a constraint that defines the very possibility of the practice of ethics. In that a merely procedural ethics founded on mutual respect of those who participate requires no more than understanding ethics as an alternative to force, such an ethic is more open to general justification than understandings of ethics that depend on particular visions of the good life and their particular orderings of benefits and banes. For further elaboration of this point, see [26], [27].

BIBLIOGRAPHY

1. Amundsen, D. A.: 1981, 'Casuistry and Professional Obligations: The Regulation of Physicians by the Court of Conscience in the Late Middle Ages' (Part I), *Transactions and Studies of the College of Physicians of Philadelphia* **3** (2), 22–29.
2. Amundsen, D. A.: 1981, 'Casuistry and Professional Obligations: The Regulation of Physicians by the Court of Conscience in the Late Middle Ages' (Part II), *Transactions and Studies of the College of Physicians of Philadelphia* **3** (30), 93–112.
3. Bacon, Sir F., 1665, *Opera Omnia*, Kampffer Matthaeaus, Frankfurt.
4. Bean, W. B.: 1982, *Walter Reed: A Biography*, University Press of Virginia, Charlottesville, Virginia.
5. Bichat, X.: 1801, *Anatomie général appliquée à la physiologie et la médicine*, 4 vols., Brisson, Gabon, Paris.
6. Bieganski, W.: 1894, *Logika Medyzyny*, Kowalewski, Warazawa.
7. Bigelow, J.: 1835, *A Discourse on Self-Limited Diseases*, Hale, Boston (delivered before the Massachusetts Medical Society, at their annual meeting 27 May 1835).
8. Blane, G.: 1819, *Elements of Medical Logik*, T. and G. Underwood, London.
9. Bonet, T.: 1679, *Sepulchretum*, Chouet Leonards, Paris.
10. Brown, B.: 1866, *On the Curability of Certain Forms of Insanity, Epilepsy, Catalepsy, and Hysteria in Females*, Hardwicke, London.
11. *Brown v. Hughes*, 94 Colo. 295, 30 P. 2d 259 (1934).

12. Carey, J. S.: 1979, 'Veterans Administration Coronary Cooperative Study', *Journal of the American Medical Association* **241**, 2791–2792.
13. *Carpenter v. Blake*, 60 Barb., 488 (N.Y., 1871).
14. Crile, G., Jr.: 1974, 'The Breast Cancer Controversy', *Transactions and Studies of the College of Physicians of Philadelphia* **41** (4), 243–253.
15. Cullen, W.: 1816, *The First Lines of the Practice of Physic*, Thomas Dobson, Philadelphia.
16. Curren, W. J.: 1969, 'Governmental Regulation of the Use of Human Subjects in Medical Research: The Approach of Two Federal Agencies', in P. A. Freund (ed.), *Experimentation with Human Subjects*, George Braziller, New York, pp. 402–454.
17. Dolan, T. M.: 1890, *Pasteur and Rabies*, Bell, London.
18. Earle, P.: 1887, *The Curability of Insanity*, Lippincott, Philadelphia.
19. Elstein, A. S.: 1979, 'Human Factors in Clinical Judgement', in H. T. Engelhardt, Jr., S. F. Spicker, and B. Towers (eds.), *Clinical Judgement*, Dordrecht, Holland, pp. 17–28.
20. Engelhardt, H. T., Jr.: 1974, 'The Disease of Masturbation; Values and the Concept of Disease', *Bulletin of the History of Medicine* **48**, 225–239.
21. Engelhardt, H. T., Jr.: 1978, 'Basic Ethical Principles in the Conduct of Biomedical and Behavioral Research Involving Human Subjects', in the National Commission for the Protection of Human Subjects of Biomedical and Behavioral Research, *The Belmont Report: Appendix Vol. I*, DHEW (OS) No. 78-0013, Government Printing Office, Washington, D.C., section 8, pp. 1–45; revised version: 1979, 'Basic Ethical Principles in the Conduct of Biomedical Research Involving Human Subjects', *Texas Reports on Biology and Medicine* **38**, 132–168.
22. Engelhardt, H. T., Jr., 1981, 'Clinical Judgement', *Metamedicine* **2**, 301–317.
23. Engelhardt, H. T., Jr.: 1981, 'The Concepts of Health and Disease', in A. Caplan, H. T. Engelhardt, Jr., and J. J. McCartney (eds.), *Concepts of Health and Disease: Interdisciplinary Perspectives*, Addison-Wesley Publishing Co., Reading, Massachusetts, pp. 31–45.
24. Engelhardt, H. T., Jr.: 1982, 'Bioethics in Pluralist Societies', *Perspectives in Biology and Medicine* **26**, pp. 64–78.
25. Engelhardt, H. T., Jr., with M. Malloy: 1982, 'Suicide and Assisting Suicide: A Critique of Legal Sanctions', *Southwestern Law Review* **36**, 1003–1037.
26. Engelhardt, H. T., Jr.: 1983, 'The Physician-Patient Relationship in a Secular Pluralist Society', in E. Shelp (ed.), *Patients and Physicians: Historical Perspective and Philosophical Inquiries*, D. Reidel Publishing Co., Dordrecht, Holland, pp. 253–256.
27. Engelhardt, H. T., Jr.: 1986, *The Foundations of Bioethics*, Oxford University Press, New York.
28. Engelhardt, H. T., Jr.: 1987, 'Introduction', in H. T. Engelhardt, Jr., and A. L. Caplan (eds.), *Scientific Controversies: Studies in the Resolution and Closure of Disputes Concerning Science and Technology*, Cambridge University Press, N.Y., pp. 1–23.
29. Eyler, J. M.: 1979, 'Statistics: A Science of Social Reform', in J. M. Eyler (ed.), *Victorian Social Medicine*, Johns Hopkins University Press, Baltimore, Maryland, pp. 13–36.

30. Farr, W.: 1885, *Vital Statistics: A Memorial Volume of Selections from the Reports and Writings of William Farr*, N. Humphries (ed.), Sanitary Institute, London.
31. Fisher, R. A.: 1925, *Statistical Methods for Research Workers*, 13th edition, Hafner, New York, 1958.
32. Fisher, R. A.: 1935, *The Design of Experiments*, 7th edition, Hafner, New York, 1960.
33. Fleck, L.: 1935, *Entstehung und Entwicklung einer wissenschaftlichen Tatsache*, Benno Schwabe, Basel; English edition: 1979, *Genesis and Development of a Scientific Fact*, T. J. Trenn and R. K. Merton (eds.), F. Bradley and T. J. Trenn (trans.), University of Chicago Press, Chicago, Illinois.
34. Foucault, M.: 1973, *The Birth of the Clinic: An Archeology of Medical Perception*, A. M. Sheridan (trans.), Random House, New York; French edition: 1963, *Naissance de la clinique. Une archéologie du regard médical*. P.U.F. Galien, Paris.
35. Gavaret, J.: 1840, *Principes généraux de statistique médicale*, Bechet jeune et Labe, Paris.
36. Gedye, J. L.: 1977, 'Simulating Clinical Judgment: An Essay in Technological Psychology', in H. T. Engelhardt, Jr., S. F. Spicker, and B. Towers (eds.), *Clinical Judgment: A Critical Appraisal*, D. Reidel Publishing Co., Dordrecht, Holland, pp. 93–114.
37. Hanson, N. R.: 1961, *Patterns of Discovery*, Cambridge University Press, Cambridge.
38. Hanson, N. R.: 1969, *Perception and Discovery*, Freeman Cooper, San Francisco, California.
39. Hill, A. B.: 1937, *Principles of Medical Statistics*, 9th edition, Oxford University Press, New York, 1971.
40. Hill, A. B.: 1963, 'Medical Ethics and Controlled Trials', *British Medical Journal* **3**, 1043–1049.
41. Hoffman, R. G. et al.: 1980, 'The Probability of Surviving Coronary Study', *Journal of the American Medical Association* **241**, 621–627.
42. Jenner, E.: 1798, *An Inquiry into the Causes and Effects of the Variolae Vaccinae*, Dawsons of Pall Mall, London.
43. Jonas, H.: 1969, 'Philosophical Reflections on Experimenting with Human Subjects', *Daedalus* **98**, 219–247.
44. Jonas, H.: 1970, 'Philosophical Reflections on Experimenting with Human Subjects', in P. A. Freund (ed.), *Experimentation with Human Subjects*, George Braziller, New York, pp. 1–31.
45. Jones, J. H.: 1981, *Bad Blood: The Tuskegee Syphilis Experiment*, Free Press, New York.
46. Katz, J.: 1971, 'Prologue – Experiments Prior to 1939', in J. Katz, *Experimentation with Human Beings*, Russell Sage Foundation, New York, pp. 284–292.
47. Katz, J.: 1978, 'Richard C. Cabot – Some Reflections on Deception and Placebos in the Practice of Medicine', *Connecticut Medicine* **42**, 199–200.
48. Louis, P. C. A.: 1835, *Recherches sur les effets de la saignée dans quelques maladies inflammatoires*, Baillière, Paris.
49. McCormick, R.: 1974, 'Proxy Consent in the Experimental Situation', *Perspectives in Biology and Medicine* **18**, pp. 2–20.
50. Morgagni, G.: 1761, *De Sedibus et causis morborum per anatomen indagatis*, Ex Typographis Remondiniana, Venice.

51. Moxley, J. H. et al.: 'Treatment of Primary Breast Cancer', *Journal of American Medical Association* **244**, 797–800.
52. Murphy, E. A.: 1977, 'Classification and Its Alternatives', in H. T. Engelhardt, S. F. Spicker, and B. Towers (eds.), *Clinical Judgment*, D. Reidel Publishing Co., Dordrecht, Holland, pp. 59–86.
53. Murphy, M. L. et al.: 1977, 'Treatment of Chronic Stable Angina', *The New England Journal of Medicine* **297**, 621–627.
54. National Commission for the Protection of Human Subjects of Biomedical and Behavioral Research: 1978, *The Belmont Report: Ethical Principles and Guidelines for the Protection of Human Subjects of Research*, HEW (OS) No. 78-0012, Government Printing Office, Washington, D.C.
55. Nozick, R.: 1974, *Anarchy, State, and Utopia*, Basic Books, New York.
56. Osterlen, F.: 1855, *Medical Logic*, G. Whitney (ed. and trans.), Sydenham Society, London.
57. Regan, L. J.: 1949, *Doctor and Patient and the Law*, 2nd edition, Mosby, St. Louis.
58. Scriven, M.: 1979, 'Clinical Judgment', in H. T. Engelhardt, Jr., S. F. Spicker, and B. Towers (eds.), *Clinical Judgment*, D. Reidel Publishing Co., Dordrecht, Holland, pp. 3–16.
59. Sydenham, T.: 1753, Preface to 'The History of Acute and Chronic Disease', in J. Swand (ed. and trans.), *The Entire Works of Thomas Sydenham*, E. Cave, London, pp. xii–xxii.
60. Sydenham, T.: 1676, *Observationes medicae circa morborum acutorum historiam et curationem*, G. Kettilby, London.
61. Urban, J. A.: 1980, 'Treatment of Primary Breast Cancer', *Journal of American Medical Association* **244**, 800–803.
62. U.S. Department of Health, Education, and Welfare: 1979, *The Breast Cancer Digest*, NIH Pub. #80–1691. Bethesda, Maryland, pp. 26–35.
63. Vrhovac, B.: 1977, 'Placebo and Its Importance in Medicine', *Journal of Clinical Pharmacology* **15** (4), 161–165.
64. Wulff, H.: 1981, *Rational Diagnosis and Treatment*, 2nd edition, Blackwell Scientific, Oxford.
65. Young, R.: 1970, *Mind, Brain and Adaptation*, Oxford University Press, Oxford.

STUART F. SPICKER

RESEARCH RISKS, RANDOMIZATION, AND RISKS TO RESEARCH: REFLECTIONS ON THE PRUDENTIAL USE OF "PILOT" TRIALS

> The two methods which we have now stated have many features of resemblance, but there are also many distinctions between them. Both are methods of *elimination*. This term . . . is well suited to express the operation, analogous to this, which has been understood since the time of Bacon to be the foundation of experimental inquiry, namely, the successive exclusion of the various circumstances which are found to accompany a phenomenon in a given instance, in order to ascertain what are those among them which can be absent consistently with the existence of the phenomenon. The Method of Agreement stands on the ground that whatever can be eliminated is not connected with the phenomenon by any law. The Method of Difference has for its foundation, that whatever cannot be eliminated is connected with the phenomenon by a law.
> Of these methods, that of Difference is more particularly a method of artificial experiment ([33], p. 256).
>
> John Stuart Mill – *A System of Logic* (1843)

I. INTRODUCTION

I should like to begin my comment and response to Professor Engelhardt by following his method, and ask you briefly to consider an episode from the history of the *acquisition* of scientific knowledge – but from a slightly different perspective and for somewhat other reasons than Dr. Engelhardt's excursion into the past. That is, we might take a moment to reflect on the compelling concerns of the past, which, over time, have led to this volume with its principal themes and which make them important enough to warrant our extensive time and energy as we speak and write to the multifarious issues germane to the general topic – the use of human beings in research [39] [42]. For these many issues have been and still are being treated not only in Israel and the United States, but throughout the world wherever support for such research is available. I turn, then, to the early 17th century.

It has always been irksome to independent thinkers, whether they be mediocre minds or persons of great genius, to have restrictions placed on them by established authority. The paradigm case in science – though given various interpretations by numerous commentators – is, of course, the complex episode between Galileo Galilei and the Roman Church during the years 1614 through 1616, and again in the disposition of 1633, wherein the Church was clearly threatened by Galilei's views and his tendency to express himself without great concern as to the sensibilities of his august audience. We might recall the fact that he was frequently criticized by his peers at Padua for offering his observations and conclusions to the lay public in Italian, rather than in the Latin of his scholarly peers. This episode perhaps predelineates the now prominent view that the scientific community is *obliged* to engage the general public in matters germane to the public's welfare and the public's health.

Restrictions by authority, however, can take a variety of forms: one may be restricted by prohibition from, and punished for, *thinking* certain thoughts; or one may be bound by censure from *expressing* certain thoughts in speech or in writing – be these true or false (recall the instructions of Father M. Segizi, Commissary of the Holy Office of the Inquisition); or, again, one may be restricted from a given practice or from *acting* in certain specified ways, e.g., preparing and executing an experiment or conducting research by implementing certain procedures, treatments, and administering new pharmacologic agents.

II. RESEARCH RISKS

Given these preliminary remarks, it would not be quite true to say, as some have, that experimentation (whether structured in rigorous designs or simply a series of anecdotal departures by physicians from standard practice) performed with respect to *inanimate* objects raises no moral problems. There are abundant examples in the history of science which, due to regulation of one form or another, raised serious ethical questions, and the Galileo episode is simply the most notorious illustration. But Galileo and others had *no* problems – other than the practical ones in conducting experiments – with the *acquisition* of knowledge. This is no longer the case. Hans Jonas, whom Dr. Engelhardt cites (and who is well known in Israel), has made the point that the manipulation of and experimentation with inanimate objects does not generally require that the investigator seek any special permission in employing his or her methods within a particular scientific design. But as soon as the *animate* world is itself the subject of inquiry, Jonas remarks, methodological innocence is lost and people are placed at risk:

> But as soon as animate, feeling beings become the subjects of experiment, as they do in the life sciences and especially in medical research, this innocence of the search for knowledge is lost and questions of conscience arise ([32], p. 2).

Hence the imposition of the Roman Church authority on Galileo was not of the sort that would serve to restrict his actual scientific work, nor his plan of research, nor the design or execution of his actual experiments, but rather it was designed to set specific *limits* on the promulgation and expression of the adequacy and truth of Copernicus' theory of heliocentric cosmography. Nor was Galileo's response to the Church authorities a plea or demand for the active support of the Copernican theory (as many incorrectly believe), as much as it was a plea or warning as to the danger which the Church would encounter in its prohibition of Copernicus' theory and the silencing of its advocates.

In any event, the so-called "Galileo affair" for our purpose simply serves to underscore the *tension* that obtained between the search for new knowledge *and* the anxieties elicited in those who honestly believed that they foresaw certain dangers which would result from scientific knowledge, but not its very acquisition. The tension today is between the search for new and useful knowledge on the one side, and *the ethical issues raised by the very inquiry itself*. The most recent illustration in our time is the spectre of risk raised by external human fertilization of living cells in laboratory culture and the reintroduction of the developing fusion-cell into the human uterus to produce an apparently normal pregnancy. Through this bioengineering technology unprecedented moral issues come to the fore, as can be gleaned from a memorandum sent September 15, 1978, to the Chairman of the Ethics Advisory Board of HEW by Joseph A. Califano, Jr., past Secretary of the U.S. Department of Health, Education and Welfare:

> ... these procedures raise serious moral and ethical questions. Does the perfection of these techniques create a potential for abuse so severe that the Federal government should not support or should strictly limit its support of the research? Can techniques of *in vitro* fertilization and transplantation of the embryo damage the resulting fetus and lead to abnormal children? Will this research lead to selective breeding, to attempt to control the

genetic makeup of offspring or to the use of "surrogate parents," where, for example, rich women might pay poor women to carry their children? ([27], pp. 155–156).

Furthermore, some even warn us that such procedures may lead to the sole use of zygotes by scientific researchers themselves for ends having nothing to do with infertility.

We should appreciate the fact that in Galileo's day the tension that existed was between a threatened institution with awesome authority – committed to the preservation of a particular cosmology – and the promulgation and *expression* of a scientific theory by those persons committed to an unbounded and totally free intellectual inquiry, e.g., Galileo.

Although the conflict and tensions continue into our own day, the nature of the conflict is really quite unique: In the United States, for example, the tension exists between a Federal institution with awesome resources and authority, *and* the everyday activities on the part of thousands of biomedical and behavioral scientists. This transformation of tensions and concerns from the early 17th century to our day was, of course, gradual and evolutionary. That is, the transformation of which I now speak culminated in the study of animate beings, not only non-human species, but *homo sapiens* as well.

One can capture the essence of this gradual transformation in Claude Bernard's *An Introduction to the Study of Experimental Medicine* [3] first published in 1865. In the process of directing his readers to the importance of the distinction between *empirical* and *experimental* inquiry (the former being in his view an inadequate, heuristic, and non-rigorous method of inquiry, the latter being experimentally rigorous and quantifiable in its results), Bernard apparently disgresses and discusses the topic of animal and human vivisection. He notes that attention to biological processes and structures requires that the investigator "operate" on the living, original, organic material entity itself. He already appreciated the fact that animal models could only fulfill a partial and limited role in the search for new knowledge regarding the human species. Hence, vivisection was for him a necessary method by means of which reliable knowledge could be acquired about the anatomical structures and physiological processes of *homo sapiens*. Bernard indirectly speaks to the ethical aspects of vivisection (there was, of course, no anesthesia at that time) and the issue of the infringement on human inviolability, but he really does not appear too troubled with offering a justification for such interventions, either.

The history of experimentation germane to human physiology is, of course, well documented since Bernard's research as the "father of modern physiology," and we need not dwell on that here. Of importance is the transition from empirical to experimental inquiry, which he advocated, and the fact that his reference to vivisection opened the way to the myriad of ethical issues in research with human subjects, these issues being couched in the very designs generated by researchers to answer the questions posed by infirmity, pain, suffering, in short, by disease. Bernard's dedication to the employment of experiments under "controlled" conditions – that is, the use of control groups, is often taken as the *sine qua non* of experimentation in the strict sense, in contrast to the daily "experiments" said to be performed by physicians with each of their patients. When we move from animal studies to the human domain, the so-called "use" of human subjects takes on a more worrisome expression and we are led to our principal question: Is there an obligation to conduct randomized clinical trials?

III. ETHICS AND RANDOMIZATION: THE PRUDENTIAL USE OF "PILOT" TRIALS

Typically, the context for this question is the appearance of a new pharmacologic agent proffered as an apparently efficacious treatment for some disease "in" some patients. The standard treatment, in many cases, is simply not efficacious; the patient is not responding especially well to it; the prognosis is not particularly encouraging, either. So how should the evaluation of the new agent proceed [26]? The RCT is typically mentioned at this time. For only this method can enable the physician/researcher to determine knowledgeably the effects of the new agent on such patients. (I think it should be understood at the outset that if we adhere dogmatically to the principle that each and every patient must be afforded the presumed benefit of any estimated advantage for one treatment over another – regardless of how slight or how uncertain that advantage may be – then all experimentation with human subjects, whether early, consecutive ("pilot") non-control-group trials, or RCTs, is vitiated.)

Although it is said that Avicenna, the Arabic physician, already called for the use of controls, replications of results, and other statistical methods ([32], p. 24), we find a clear expression of the critical randomization canon in Mill's *A System of Logic*. In his "Second Canon," Mill formulates the Method of Difference:

If an instance in which the phenomenon under investigation occurs, and an instance in which it does not occur, have every circumstance in common save one, that one occurring only in the former; the circumstance in which alone the two instances differ is the effect, or the cause, or an indispensable part of the cause, of the phenomenon . . . ([33], p. 256).

And a few pages later he concludes: "It thus appears to be by the Method of Difference alone that we can ever, in the way of direct experience, arrive with certainty at causes" ([33], p. 258).

This Canon, of course, captures the very essence of the physician/researcher's professional mandate and has, in fact, been the keystone for those whose lives are dedicated to the scientific enterprise.

On November 6, 1973, Dr. Thomas C. Chalmers[1] addressed the members of the National Eye Institute Workshop in Washington, D.C. His topic and subsequently published paper was the "Ethical Aspects of Clinical Trials." In his opening lines, a self-confession caught my eye (as Dr. Chalmers rarely, if ever, included a personal note in his extensive list of publications): he remarked that" . . . I treated many patients incorrectly because I did not know about controlled clinical trials, and that is why I decided to devote my life to convincing students and physicians of the crucial importance of randomized controlled trials" ([7], p. 753).[2]

This candid comment reveals this physician/researcher's dedication to what he terms "ethical" medical action ([19], p. 335). Here, as is common to note among researchers, the term "ethical" needs qualification, but usually connotes attention to the following maxims: (1) do not gratuitously place patients/subjects at risk; do not enter or continue them in a study that in principle can yield no meaningful clinical results ([24] p. 694); (2) do not harm the patients/subjects; do not risk them beyond "minimal risks" by entering them into a randomized clinical trial (RCT) – in which case each individual has a fifty percent chance of being in either the new drug (experimental) group or the "standard treatment" (control) group; (3) do not continue a study in which patients/subjects are taking an ever-increasing risk of injury or harm, including the danger of death.[3] Furthermore, (4) do not deprive patients of treatments known to be efficacious; (5) generally speaking, protect the patient from toxicity, i.e., the over-zealous physician may be biased with regard to "known" efficacy/toxicity ratios ([19], p. 333); (6) only employ no-treatment control groups when no effective therapy at all exists ([2], p. 795). Finally, (7) it is generally not protective of them "to be randomly assigning patients to a placebo" ([7], p. 754).

On the basis of these maxims, it is clear that we are focussing not on healthy subjects – who might conceivably participate in non-therapeutic protocols or, if prevention is one's goal, even in therapeutic protocols – but on the sick patient/subject, who is a potential candidate for a therapeutic protocol. Here I shall not pursue the question of the moral permissibility or impermissibility of employing sick patients in non-therapeutic trials, even with their expressed consent.[4] Suffice to say, I tend to agree with Hans Jonas and Thomas Chalmers, and not with Dr. Engelhardt for whom such protocols are acceptable since they "support the good they [the sick] have endorsed" ([22], p. 135). Dr. Chalmers states his view quite straightforwardly. Using sick patients in non-therapeutic studies is generally unacceptable, even if they are "true volunteers" and "fully understand" the study and their rôle in it. "The only condition for getting a patient to enter a study . . . is that he has a reasonable opportunity of receiving benefit, and if he is only entering a study in order to prove something for all future patients, he should fully understand that he is a true volunteer. I think you need careful peer review before you do that" ([7], p. 757).

These general moral maxims are all germane to the most fundamental and most practical of criteria by which one may judge a treatment: "the general well-being of a patient" ([28], p. 125). Assuming this principle for the present, I should like to be clear about what I do not disagree with in order to be very clear on where I disagree with the clearly erroneous view concerning the "ethical/unethical" aspects of RCTs.

First, I do not deny that, when properly conducted, RCTs effectively eliminate observer and other bias; I do not deny that clinical, randomized, quadruple-blinded[5] studies are feasible; I do not deny that so-called "historical controls" are fraught with difficulties – including the highly questionable results of these trials; they typically (though not intrinsically) reflect biased patient selection in their recourse to patient records and the studies published in the literature ([37], p. 233).[6] Most importantly, I follow J. S. Mill in holding that for acquiring *knowledge of causes* there is really no substitute for the Method of Difference and the technique known as the RCT.

Having partially qualified my position, let me now summarize the debate:

In a letter to the editor of *The New England Journal of Medicine* (1981) three faculty members of the Mt. Sinai School of Medicine in New York, including Dr. Chalmers, charged the investigators of two anecdotal, "uncontrolled and unblinded" trials with a violation of the "basic principles of scientific inquiry," with the implication that the study, since it was uncontrolled, ought not to have been undertaken in the first place. The Mt. Sinai doctors also argued that, "If they are convinced of the effectiveness of these drugs, can they ethically do blinded and randomized trials . . . when to do so would mean withholding a 'proved' drug from a patient with a life-threatening disease?" ([36], p. 1067).

Notwithstanding the Mt. Sinai doctors' uncompromising concern with the welfare of the sick patients who participated in the study at great risk to themselves, there are *two* important questions that must be distinguished and addressed: First, should uncontrolled ("no-control group") clinical trials ever be performed when a scientifically quadruple-blinded controlled RCT is feasible? Secondly, is it possible for investigators, who are uncompromisingly concerned with their patients' safety and well-being, *ethically* to conduct an uncontrolled trial, especially when they have some good reason to *believe* on the basis of anecdotal and other very limited evidence that the new drug or treatment *may* be better than the standard and reasonably well-known drug regimen?

A careful reading of the Mt. Sinai physicians' letter reveals that they tend to beg the question at issue, since they somewhat unfairly attribute a very "fixed belief" to the investigators, by claiming that the latter are "convinced" of the new drug's efficacy. *But the investigators were in fact not convinced in the sense of knowing; they were simply strongly hoping and believing that the drug under investigation would eventually, after an RCT, prove to be more efficacious than the standard treatment.* That this was in fact an incorrect *assumption* by the doctors is made clear by their remark that such a controlled clinical trial "would mean withholding a 'proved' drug from a patient with a life-threatening disease." By placing the word 'proved' in quotation marks, the position of the researchers is actually sustained: that the drug in question, though *believed* to be efficacious, was not yet proved to be so. The answer to the first question, then, is "yes". The *process* of scientific inquiry warrants the use of uncontrolled, consecutive, clinical "pilot" trials, and too quick an application of an RCT of the "first patient"[7] would be, in many but not all cases, ethically *unjustified*. I therefore take issue with Dr. Chalmers' claim that "whenever there is uncertainty about proper diagnosis and therapy, scientific [randomized, controlled] clinical trials are the most ethical way to benefit both the individual patient and all others" ([19], p.

325). The answer to the second question is also "yes". For one's belief about a new drug's efficacy in no way bars a physician/researcher from conducting an RCT at a later time, having initiated and completed a series of pilot studies ([7], p. 755; [8], p. 1036).

This point requires careful elaboration: The best way to resolve the apparent debate, which otherwise may continue indefinitely – whether or not the pilot study should give way in all cases to the RCT – is to appreciate the fact that there are intermediate phases *between* these two extremes. By recognizing this we can avoid the conclusion that the "first patient" should be randomized. At this point statistical sophistication may well spell the resolution of the dilemma, and also serve to resolve the apparent *ethical* quandary: Whether it is morally permissible or not to allow a small number of patients to receive a new drug which the physician/researcher *believes* will prove more efficacious than the present or standard treatment.[8]

Again, there is no dispute about the fact that the properly conducted RCT is far more likely to reveal the "final evidence", whereas the process of gathering preliminary information is in most cases unlikely to yield fully adequate evidence in order to enable us to fix our belief, to know with certainty the efficacy of the new pharmacologic substance. What is at stake here is the relation of *belief* to *knowledge* – but knowledge is simply *justified belief*. In a letter of response to another physician, Dr. Chalmers correctly distinguishes the term "randomization" from the term "controlled clinical trial". He points out that one should not equate or employ these terms interchangeably. He later suggests that "There is no reason why randomization should not be an essential part of . . . the 'precontrolled trial'" ([9], p. 70). Thus, he continues here and elsewhere ([7], p. 756), "randomization does not even have to be between patients, but could be between doses or between times of administration, or between active and inactive medication." Thus he concludes that " . . . early randomization is both better statistically and more ethical", that is, less risky to patients. One might wonder whether Dr. Chalmers is intellectually more desperate here. Can one really gain knowledge, justify one's belief in a drug's efficacy, for example, by *randomizing* the first patient's drug regimen? It is well known that the subject of the study would himself be undergoing changes that "bias" the on-going study. Or, put another way, how is this new suggestion any different from a carefully planned and monitored consecutive pilot trial, or set of trials, with an ever-increasing number of patients, prior to the running of an RCT? To employ the method of "randomization of the doses", or "randomization of the times of administration", or "randomization of active and inactive medication" is, in truth, to *fail* to apply the principle of *ceteris paribus* ("all other things being held constant"), since by definition that is not quite possible in these early, consecutive non-comparative pilot trials ([40]), p. 263).[9] Even Dr. Chalmers says that "the randomization process" must be "pristine", which means contrived according to strict standards, and is not a mere matter of chance. (See [11], p. 453). In short, Dr. Chalmers' letter of February, 1975 to *The American Statistician* ([9], p. 70) contains the seeds of the refutation of two of his own earlier arguments:

First, the optimistic belief that a given new drug is more efficacious than the present standard treatment *is no ethical barrier* to properly conducted pilot trials with slowly incremental numbers of patients, but perhaps no fewer than six. Secondly, non-randomized studies are more often than not actually *ethically obligatory* prior to the establishment of an RCT, and therefore one need not have recourse to randomization[10] of the precontrol trial patients. Quite often randomization is neither possible nor proper and the physician/researcher, like most of us, has to live with the slow though now very sophisticated processes required by the *acquisition* of knowledge, which progresses from patient to patient where 50:50 odds may not be the best one can conjure. One might ask whether there is at present adequate evidence to allow one to conclude that patients, when considering a decision to participate or not in an RCT, would in fact prefer a 50:50 chance of receiving a new drug when it might be the case – based on statistical calculations of pilot trials – that there is a 60:40 or better chance that the new drug will be better than the standard treatment. In short, statisticians and research designers are only now exploring the psychology of patient self-evaluations of risk/benefit decisions. What Dr. Chalmers terms the most "rational" decision is, here, open to further evidence. I therefore do deny that a series of pilot, consecutive trials with patients who are not improving from standard therapy should be eliminated from medical methodology, whereas Dr. Chalmers rejects these pilot studies ([15], p. 1182; [13], p. 107; [6], p. 5). I deny that observer bias is detrimental in such pilot studies; I deny that these pilot trials are unethical. As Dr. E. A. Gehan has pointed out, "there is a case for selected as opposed to a randomized control group in certain circumstance" ([25], p. 198): the *preliminary estimate* of the effectiveness of the proposed, new regimen of treatment.

To be sure, these early trials must not involve hundreds of patients ([28], p. 121–2). The point is that a physician's confidence plays an extremely important rôle in the very earliest estimates about the efficacy of a new drug, and that confidence level is open to statistical procedures that can be applied in a very careful fashion. As Julian Simon remarks: given our "massive ignorance" of a new drug's efficacy, it requires "a series of rough trial-and-error stages." Only after this series "should we be ready to run controlled experiments . . ." ([40], pp. 268–269).

In fairness it should be mentioned that Dr. Chalmers has pointed out that the "scientific objections" to randomizing the first patient are "legitimate" ([6], pp. 6–7). It is the "ethical standpoint" that most concerns him, and thus he maintains that early, consecutive trials "would be wholly unacceptable." As I have tried to show, Dr. Chalmers' claim that "Randomization from the beginning, with truly informed consent, is the only *ethical* way to begin the exploration of new therapies", is not a sound maxim. Although the informed consent is demanded, the randomization from the beginning is *not* a necessary condition for an ethical study.

A review of the literature reveals that I am not the first to take up the challenge posed by Dr. Chalmers' professional, life-long conviction. Dr. J. S. Hagans ([9], p. 70) and the statistician, Paul Meier, have already challenged this view. Meier notes that we might at times employ pilot, informal, therapeutic innovations without any ethical compromises regarding the patient's welfare ([32], p. 19). Without pre-RCT "baseline characteristics" of a new drug, a classically trained physician would properly view the RCT of the first patient as a "monstrous act of ethical irresponsibility" ([32], p. 10). The everpresent danger, then, is that the RCT can become a "mischievous ritual" deflecting our concern from the patient who is already unresponsive to or showing no significant improvement from the standard treatment. And, hence, the RCT is not the only way to the truth; only the last way. The fact most ignored by commentators on the concept of induction and in Mill's Second Canon on the Method of Difference is the fact that he was not thinking of determining causes in *living* beings, but only with regard to *inanimate*, natural processes. In the final analysis, when all the circumstances pertinent to each unique patient and the earlier information gleaned from the animal studies – including the knowledge of the important patient characteristics influencing prognosis – are taken into account, we may well be led to undertake the rational and ethical decision to postpone the RCT in favor of early, consecutive, pilot trials. In the end, the proper use of such pilot trials can hopefully give us *approximate answers* to the right questions; this may well portend a better state of affairs than seeking prematurely an exact answer to a merely *approximate problem*.

IV. RISKS TO RESEARCH

Please permit me to close with a reference to the School of Pythagoras, 6th century B.C. In his *Metaphysica*, Aristotle reviews the history of philosophy until his time; he remarks that

> ... the so-called Pythagoreans, who were the first to take up mathematics, not only advanced this study, but also having been brought up in it they thought its principles were the principles of all things, since of these principles numbers are by nature the first, and in numbers they seemed to see many resemblances to the things that exist and come into being. Since then all other things seemed in their whole nature to be modelled on numbers, and numbers seemed to be the first things in the whole of nature, they supposed the elements of numbers to be the elements of things ... ([1], A, 5, 985 b23–986 a2).

For the Pythagoreans, then, "number was the substance of all things" ([1], A, 5, 987 a19) and "the things themselves are Numbers" (whereas for Plato "the Numbers exist apart from sensible things) ([1], A, 6, 987 b27–29). The essential reality in Pythagorean metaphysics has dominated the thinking of Western science, especially since the beginning of its renaissance in the 17th century and into our own day. Of course we have come quite far (however one defines "progress" through these centuries), but now there is a nuance of meaning that the Pythagoreans could not have foreseen.[11] From the notion that number expresses quintessential reality, we have come to emphasize and pay homage to the *Number* ('N') of subjects in a randomized, controlled clinical trial.

As I have raised the spectre of the "minimum N" necessary for a reliable RCT, I shall conclude with a few remarks on this oft-neglected topic. The issue is the risk to research itself, wherein researchers are in danger of not having adequate subjects to conduct an RCT properly ([17], p. 448). And here I only mention the materially troubling fact, studied very carefully by Dr. Chalmers and others, that far too many RCTs are conducted without an adequate 'N'. I shall simply assume that this present state of affairs will be ameliorated, because men like Dr. Chalmers have spent their lives teaching medical students and other physicians the necessary and sufficient conditions for conducting reliable RCTs, whose results are the best we can obtain in the long run, presuming we make the long-time effort.

In a very important study, J. E. Freiman, Chalmers, and others reexamined some 71 "negative" RCTs to determine if the researchers had employed a large enough sample 'N' to give a high probability of detecting a 25 percent and 50 percent therapeutic improvement in the patient-subjects. In an astonishing "Discussion" section, the authors note that "[a]nalysis of the 71 'negative' randomized control trials shows that the investigators often work with sample sizes too small to offer a reasonable chance of successfully rejecting the null hypotheses in favor of the treatment" ([24], p. 694). That is, "of the 71 so-called negative papers in which data were recalculated, only 14 had sufficient numbers to warrant the negative statement of the authors, i.e., to rule out a 25 percent improvement" ([19], p. 332).

In the final analysis, the biostatistician is the person responsible for "the estimates of numbers of patients required" ([6], p. 7). This estimate of the necessary, "minimum N" enables the researcher to conclude with confidence that it is probable, at a selected "p value", that an observed difference was not due to *chance*, and therefore it is possible to make reasonably sure that, according to accepted confidence limits, a clinically meaningful *difference* was not being missed by the investigator.

From the contemporary researcher's perspective "all *is* number", and valid conclusions hang on the appropriate sample "N". For statistically significant results can only be obtained when the sample "N" (control/experimental groups) is the minimally necessary number of participants [24]. But the statistical determination and actual achievement of this goal – with the voluntary, informed, consensual participation of persons in the study – is a primary concern of every competent physician/researcher. And this goal, thanks to the ever-demanding requirements and *regulations* on researchers, is in constant danger of being thwarted and thereby serves to risk the research enterprise as such. The best way to show this is by recourse to a whimsical illustration – for humor often contains the seed of the serious.

In certain circles there is discussion of what is called Lasagna's Law, a law which explains why clinical trials come to grief. It goes like this: prior to the beginning of a trial, the number of subjects ("N") available to enter the study is adequate. When the trial actually begins, the "N" is a fraction of what it was said to be before the trial began. As soon as the trial is terminated the pre-trial level returns and the "N" is again adequate for the trial [26].

How to account for this mysterious phenomenon? Did the subjects in fact disappear and reappear? Or is it simply that the "ethical" or "safety" or "risk free" demands of the protocol require a burdensome list of exclusions which, in the end, serve to invalidate the study?[12] In trials to determine the efficacy of new drugs, for example, women of childbearing potential, children, the very elderly, those incapable of giving an informed consent, those too ill to participate, may all be excluded. Did the subjects pass their laboratory fitness test? Are they abnormal? Was there a laboratory error? Moreover, did they withdraw during the early phase of the longitudinal study for good or inadequate reasons? Who falls outside the study's pre-determined confidence limits? Who has a chronic disease and therefore will, if counted, "contaminate" the project? Did the researcher's detailed explanation make the previously consenting participants too anxious – so they chose to withdraw?

There are, of course, ways that researchers can, by reducing the "N", try to defeat this process, but can they ethically conduct a study when it is known beforehand that the results can reveal no new knowledge because the project was doomed from the beginning, since the required, minimum "N" was not obtained? Should we encourage researchers to change (lower?) the minimum consent and numerical standards for persons to enter the trial? Should we encourage the abandonment of the randomized clinical trial in favor of "pilot," consecutive trials and anecdotal evidence?

In short, there are today researchers who maintain that the risks *to* research are beginning to equal the risks *from* research, and a paradoxical situation thus faces us. For even under the best sample conditions, the truly randomized collection of participants is never a truly representative selection. If we allow numerous, additional exclusions things may go from bad to worse to impossible.

Lasagna's Law has, of course, a wonderful and promising corollary – the Prophylactic corollary – for notwithstanding the fact that no research results of any import can any longer be gleaned from clinical trials, we could take comfort in the fact that serious disease will fall to an all-time low. This will result when researchers, following the prescriptions for public health studies, institute clinical trials of the most deadly diseases. Lasagna's Law necessarily leads to the analytic truth that if "N" is zero, then no one can suffer disease.[13]

School of Medicine
University of Connecticut Health Center
Farmington, Connecticut, U.S.A.

NOTES

[1] Thomas C. Chalmers, M.D., is Past-President of the Mount Sinai Medical Center, and Dean, Mount Sinai School of Medicine, New York City.

[2] Other eminent medical methodologists, like Max Hamilton, have shared Dr. Chalmers' view of the importance of RCTs ([28], pp. 121–22).

[3] It is, of course, frequently ethically required to discontinue a trial when it is clear that patients are being harmed either by the new treatment or some unanticipated adverse reaction. The famous Thalidomide study was interrupted and, of course, this vitiates the drawing of any sound scientific *conclusions*, though surely a sound moral *decision* was made ([32], pp. 3, 16–17, 21). Also see [19], pp. 330–331; [18], p. 34; [17], p. 451 and [11], p. 453.

[4] I shall not take up the serious matter of the requirement of informed consent, but rather presume it in the case of pilot studies and RCTs. See [43], [21], [23], [35].

[5] Quadruple blinding refers to: 1) the randomization process itself; 2) the physician does not know what the patient receives; 3) the patient does not know what he/she receives; 4) the physician does not know the on-going results of the study ([18], p. 31).

[6] Also see [16], p. 1293; [6], p. 6; [14], p. 233.

[7] Dr. Chalmers' stand on RCTs for the "first patient" was asserted as early as 1961 by Max Hamilton. See [28], pp. 121–122.

[8] In fairness to Dr. Chalmers' view, we call attention to Max Hamilton's remarks:

"The real problem is not whether the physician should refrain from giving some of his patients a new and untried drug, but whether he is ethically justified in *giving* it to some of his patients. Again and again, as soon as a new treatment is introduced, within a short period I have seen reports of its effect, it having been administered to scores and even hundreds of patients. Those reports are not controlled and they are generally very enthusiastic over their remarkable results!" ([28], pp. 121–22)

[9] As Max Hamilton says, paraphrasing J. S. Mill mentioned in our opening remarks, "we obtain our control series by trying to find subjects who are like those under observation in every respect except the one under consideration" ([28], p. 117).

[10] It is surprising how few writers on issues in human experimentation correctly distinguish *randomization* from mere *chance occurrence*. "A sample is random if every member of the universe stands the same chance of being included in it" ([40], p. 39). We should note that human subjects are not typically sorted into groups the way, for example, a ball falls in any one slot of a perfectly balanced roulette wheel. Furthermore, a random

assignment may not be *representative*, since even patients with a similar type of carcinoma are different in many, and sometimes essential, respects. So researchers must seek a perfect representation of the population in question.

[11] A significant step had already been taken in the 3rd Century B.C. by Archimedes in his truly experimental work with floating bodies, e.g., the famous story of the problem of the crown ([29], pp. 92–94).

[12] See [7], p. 753.

[13] I wish to thank my colleague and friend, Tris Engelhardt, for his careful critique of this essay. Without his scrupulous sense of argument and detail, I could not have managed to present it to a wider audience. For the remaining difficulties I, of course, am fully responsible.

BIBLIOGRAPHY

1. Aristotle: 1908, *The Works of Aristotle*, (trans.) W. D. Ross, Clarendon Press, Oxford, England.
2. Baum, M. L., Anish, D. S., Chalmers, T. C., *et al.*: 1981, 'A Survey of Clinical Trials of Antibiotic Prophylaxis in Colon Surgery: Evidence Against Further Use of No-Treatment Controls', *The New England Journal of Medicine* **305**, 795–799.
3. Bernard, Claude: 1957, *Introduction to the Study of Experimental Medicine* (1865), Dover Publications, New York. See Part Two, Chapter II, Section III.
4. Bonchek, L. I.: 1979, 'Are Randomized Trials Appropriate for Evaluating New Operations?', *New England Journal of Medicine* **301** (1), 44–45.
5. Chalmers, T. C.: 1972, 'Editorial – Randomization and Coronary Artery Surgery', *Annals of Thoracic Surgery* **14** (3), 323–327.
6. Chalmers, T. C.: 1976, 'Randomized Clinical Trials in Surgery', in R. L. Varro and J. P. Delaney (eds.), *Controversy in Surgery*, W. B. Saunders Co., Philadelphia, Pennsylvania, pp. 3–11.
7. Chalmers, T. C.: 1975, 'Ethical Aspects of Clinical Trials', *American Journal of Ophthalmology* **79** (5), 753–758.
8. Chalmers, T. C.: 1975, 'Randomization of the First Patient', *Medical Clinics of North America* **59** (4), 1035–1038.
9. Chalmers, T. C.: 1975, 'Reply – Randomization in Clinical Trials', *The American Statistician* **92** (1), 70.
10. Chalmers, T. C.: 1975, 'The Impact of Clinical Trials on the Practice of Medicine', *Mount Sinai Journal of Medicine* **41**, 753–759.
11. Chalmers, T. C.: 1976, 'Randomized Controlled Clinical Trials in Diseases of the Liver', in H. Popper and F. Schaffner (eds.), *Progress in Liver Diseases*, Vol. 5, Grune and Stratton, New York, especially Chapter 27, pp. 450–455.
12. Chalmers, T. C. and Discussants: 1976, 'How to Turn Off an Experiment', in J. D. Cooper and H. D. Ley (eds.), *Ethical Safeguards in Research on Humans*, Interdisciplinary Communications Associates, Washington, D.C., pp. 119–143.
13. Chalmers, T. C.: 1977, 'Letter – Randomize the First Patient!', *The New England Journal of Medicine* **296** (2), 107.
14. Chalmers, T. C., *et al.*: 1978, 'In Defense of the VA Randomized Control Trial of Coronary Artery Surgery', *Clinical Research* **26** (4), 230–235.

15. Chalmers, T. C.: 1979, 'Letter – Randomized Clinical Trials in Surgery', *The New England Journal of Medicine* **301** (21), 1182.
16. Chalmers, T. C. and Schroeder, B.: 1979, 'Letter – Controls in *Journal* Articles', *The New England Journal of Medicine* **301** (23), 1293.
17. Chalmers, T. C., Sacks, H. S. and Schroeder, B. J.: 1981, 'Critique of Methods and Timing', in P. W. Dykes and M. R. B. Keighley (eds.), *Gastrointestinal Haemorrhage*, Wright P. S. G, Bristol, England, pp. 448–454.
18. Chalmers, T. C. and Smith, H., *et al.*: 1981, 'A Method for Assessing the Quality of a Randomized Control Trial', *Controlled Clinical Trials* **2**, 31–49.
19. Chalmers, T. C.: 1981, 'The Clinical Trial', *Milbank Memorial Fund Quarterly: Health and Society* **59** (5), 324–338.
20. Chalmers, T. C.: 1982, 'Discussion – Combinations of Data from Randomized Control Trials', *Biometrics* **38** (1), 150–153.
21. Curran W. J.: 1979, 'Reasonableness and Randomization in Clinical Trials: Fundamental Law and Governmental Regulation', *New England Journal of Medicine* **300** (22), 1273–1275.
22. Engelhardt, H. T.: 1988, 'Diagnosing Well and Treating Prudently', in this volume, pp. 123–141.
23. Fost, N.: 1979, 'Consent as a Barrier to Research', *The New England Journal of Medicine* **300** (22), 1272–1273.
24. Freiman, J. A., Chalmers, T. C., Smith, H. and Kuebler, R. R.: 1978, 'The Importance of Beta, the Type II Error and Sample Size in the Design and Interpretation of the Randomized Control Trial: Survey of 71 "Negative" Trials', *The New England Journal of Medicine* **299** (13), 690–694.
25. Gehan, E. A. and Freireich, E. J.: 1974, 'Non-Randomized Controls in Cancer Clinical Trials', *The New England Journal of Medicine* **290** (4), 198–203.
26. Gorringe, J. A. L.: 1970, 'Initial Preparation for Clinical Trials', in E. L. Harris and J. D. Fitzgerald (eds.), *Principles and Practice of Clinical Trials*, Livingston, Edinburgh, pp. 41–46.
27. Grobstein, Clifford: 1981, *From Chance to Purpose: An Appraisal of External Human Fertilization*, Addison-Wesley Publishing Co., Reading, Massachusetts.
28. Hamilton, M.: 1974, *Lectures on the Methodology of Clinical Research* (1961), Churchill Livingstone, Edinburgh and London. Especially Lecture 10, pp. 113–123.
29. Heath, T.: 1921, *A History of Greek Mathematics*, Vol. II, Clarendon Press, Oxford, England. Especially Chapter XIII.
30. Hollenberg, N. K., *et al.*: 1981, 'Letter – Are Uncontrolled Clinical Studies Ever Justified?', *The New England Journal of Medicine* **303** (18), 1067.
31. Jonas, H.: 1969, 'Philosophical Reflections of Experimenting with Human Subjects', in P. A. Freund (ed.), *Experimentation with Human Subjects*, George Braziller, New York, pp. 1–31.
32. Meier, Paul: 1974, 'Statistics and Medical Experimentation', (unpublished paper) delivered as the ENAR Presidential Invited Address, Tallahassee, Florida, March 22.
33. Mill, J. S.: 1959, *A System of Logic: Ratiocinative and Inductive* (1843), Longmans, Green and Co., London. See especially Book III, Chapter VIII, 'Of the Four Methods of Experimental Inquiry,' pp. 253–266.
34. National Commission for the Protection of Human Subjects of Biomedical and

Behavioral Research, *The Belmont Report: Ethical Principles and Guidelines for the Protection of Human Subjects of Research* (1978), DHEW Publication No. (OS), No. 78–0012, U.S. Government Printing Office, Washington, D.C.
35. Relman, A. S.: 1979, 'The Ethics of Randomized Clinical Trials', *The New England Journal of Medicine* **300** (22), 1972.
36. Sacks, H., *et al.*: 1981, 'Letter – Are Uncontrolled Clinical Studies Ever Justified?', *The New England Journal of Medicine* **303** (18), 1067.
37. Sacks, H., Chalmers, T. C. and Smith H.: 1982, 'Randomized versus Historical Controls for Clinical Trials', *American Journal of Medicine* **72**, 233–240.
38. Shaw, L. W. and Chalmers, T. C.: 1970, 'Ethics in Cooperative Clinical Trials', *Annals of the New York Academy of Science* **169**, 487–495.
39. Shimkin, M. B.: 1953, 'The Problem of Experimentation on Human Beings: The Research Worker's Point of View', *Science* **117** (3035), 205–207.
40. Simon, J. L.: 1969, *Basic Research Methods in Social Science*, Random House, New York. See Chapters 3 and 25.
41. Spodick, D. H.: 1975, 'Letter – "Ethics Shock" and Randomized Trials', *The New England Journal of Medicine* **292** (12), 653.
42. Wolfensberger, W.: 1967, 'Ethical Issues in Research with Human Subjects', *Science* **155**, 47–51.
43. Zelen, M.: 1979, 'A New Design for Randomized Clinical Trials', *The New England Journal of Medicine* **300** (22), 1242–1245.

ANNE M. FAGOT-LARGEAULT

EPISTEMOLOGICAL PRESUPPOSITIONS INVOLVED IN THE PROGRAMS OF HUMAN RESEARCH

I. THE ETHICS OF KNOWLEDGE AND THE ETHICS OF BENEFICENCE

The present paper addresses the question whether there are areas of conflict between currently accepted standards of scientific research in the biomedical and social sciences and ethical standards implying a respect for human beings. One may view it as a conflict between two ethics: an "ethics of knowledge" (objective truth is the highest value) and an "ethics of beneficence" (one should seek the good for mankind); or as a conflict between two conceptions of science: "pure" science, and "other" science.

The source of the problem lies in an ambiguity about the finality of scientific research. Some research is supposedly conducted for the good of someone (individuals? society? the human species?); some research is supposedly conducted in the interest of "pure" science (or the speculative or intellectual interests of researchers). Contrary to what Claude Bernard asserted somewhat hastily, the interests of science do not necessarily coincide with the good of humanity, or some community, apart from the community of researchers, perhaps. Speaking of biomedical disciplines, Henry Beecher alluded to the "unfortunate separation between the interests of science and the interests of the patient" ([3], p. 1355). It is by no means obvious that scientific and humanitarian concerns will always naturally agree.

Traditionally, the interests of science were said to be subordinated to the interests of individual persons, in countries of Roman–Napoleonic law, or subordinated to the interests of society, in Scandinavian or English-speaking countries. "The experiment should be such as to yield fruitful results for the good of society", according to the Nuremberg Code of 1946. "Science is not the highest value to which all orders of values . . . should be subordinated," said Pope Pius XII in 1952 ([3], p. 1354). And Claude Bernard observed:

It is our duty and our right to perform an experiment on man whenever it can save his life, cure him or gain him some personal benefit. The principle of medical and surgical morality, therefore, consists in never performing on man an experiment which might be harmful to him to any extent, even though the result might be highly advantageous to science, *i.e.*, to the health of others. But performing experiments and operations exclusively from the point of view of the patient's own advantage does not prevent their turning out profitably to science . . . [4].

Recently, however, the situation has become more equivocal. The regulations issued in the U.S.A. in 1981 by the D.H.H.S. and the F.D.A. have seemed to imply that research physicians are now to some extent justified in abandoning the Hippocratic rule to "serve only the patient", and that some research involving minor risks for subjects is justified by the importance of the knowledge to be gained. Respect for persons seems to some extent subordinated to considerations of the intrinsic interest of research. At least one may seek a compromise between the two.

The current situation in France, for example, is typical of such an ambiguous and transitional stage. Research in the field of biomedicine has been presumed therapeutic. "Pure" research on healthy or ill human beings has had no legal status; it has been supposed not to exist. In fact it is done, to a large extent on the sick or disabled often under some therapeutic pretence, to a lesser extent on healthy volunteers because lawyers are reticent, there is no insurance coverage, and French jurisprudence is dissuasive. In 1980 the Paris Court of Appeal pronounced a judgment which rendered phase 1 and 2, and even phase 4, of clinical trials problematic, for "human beings should be protected against so-called exploratory clinical trials, the hazardous character of which can't be ignored" [7]. Paradoxically, Phase 3 trials are marginally licit, since a decree of 1975 rendered them mandatory in the procedure preliminary to a new drug's authorization. In fact, there is a latent feeling of uneasiness among French medical doctors. Their Code of Deontology of 1979 stipulates that they should never expose any patient to an "unjustified risk" ([6], Art. 18) and that new therapies are to be used only if there is a direct benefit to the person ([6], Art. 19). Dr. Pierre Royer, a pediatrician at the Necker-Enfants-Malades Hospital in Paris, points out that the strict principle has actually been stretched over the years. In the case of children, for example, it was admitted that an experimental procedure may be used if it is potentially useful for the child (*e.g.*, administration of a new antibiotic to treat purulent meningitis), then if it is potentially useful for the family (*e.g.*, investigation of an hereditary disease to allow genetic counseling and prevention of another disabled child), then if it is potentially useful for the community at large (*e.g.*, vaccine trials), then if it is potentially useful for science and/or the human species (*e.g.*, physiological studies). There is a certain *hypocrisy*, says Royer, in thus stretching the principle ([19], p. 4090). Most official texts go on to say what was said at the First International Congress of Medical Ethics in Paris, 1955: "No investigation may take place in the interest of science if it is useless for the patient himself" [22]. The World Medical Association proclaims: "Concern for the interests of the subject must always prevail over the interest of science and society" [25]; and "The doctor's fundamental role is to alleviate the distress of his or her fellow men, and no motive whether personal, collective or political shall prevail against this higher purpose" [26]. Yet it is known that, in practice, scientific needs sometimes overcome humanitarian considerations, and some worry that such practices shall soon be legalized.

This has been a subject of concern for philosophers ([15], pp. 139 ff.; [21], p. 10), medical doctors, lawyers and journalists ([15], pp. 359 ff.). Journalists have denounced the deceit of the official research ethics. Lawyers worry that the implicit contract between physicians and patients is being violated. In clinical trials, medical doctors have been found practising an "ethical adjustment of randomization", *e.g.*, by transilluminating envelopes that contained the names of treatments, in order to give some particular patients what they thought to be a more desirable treatment; more honestly, some medical practitioners have refused to participate in blind or double-blind trials. Philosophers suspect that one is currently trying to make ethics agree with scientific needs, rather than trying to make scientific habits comply with ethics. Ethical principles may evolve indeed, but there are no "absolute" scientific standards, either! Are there any imperatives of scientific research that justify that traditional ethical standards be stretched and twisted?

I do believe that over the past two centuries the imperatives of scientific research have led us to an imperceptible modification of accepted ethical standards. One must wonder whether the recent spreading of IRB's and ethical commissions is not meant to consecrate practices hitherto considered immoral and ignored as marginal, and restore a consensus of good "enlightened" conscience at a time when such practices have become too widespread to be readily ignored. Admittedly the ethical conduct of research implies and requires good scientific methodology. (But is there only one methodological choice scientifically acceptable in each case?) On the other hand, it would be naive to think that "bad ethics makes bad science"; nor does a good scientific methodology imply ethical procedures. (Imagine some experiments performed on animals with perfect scientific rigor, which could be carried out on human subjects with the same rigor.)

In the following I shall examine some of the current presuppositions of research on human subjects. I intend to show that there may be a fundamental difference between research conducted primarily for the good of someone ("pragmatic"), and research conducted for "purely scientific" motives ("explanatory"); that this implies a difference in the scientific strategies; that the different scientific approach called for by the "ethics of benevolence" is not necessarily a less scientific approach than the one implied by the "ethics of knowledge"; further, that most research is a mixture of these two approaches, – but that mixed strategies can be worse than pure strategies, particularly if the ambiguity is not clearly stated and one oscillates from one point of view to the other for opportunist motives; finally, that lay people can easily adhere to the imperatives of pragmatic research, but that few people would knowingly risk their health or their life for the sake of "science", unless they believed that such scientific research is beneficial in the pragmatic sense.

2. THE PRESUPPOSITION THAT ONE SHOULD DO RESEARCH ON HUMAN BEINGS

The first presupposition, which is basic, is that one should do research on human beings, *i.e.*, seek scientific progress in the realm of the biomedical and social sciences; that the search for knowledge is valuable, and that in some circumstances its value should take precedence over some other human values.

Example: a 35-year-old woman with a rare syndrome – a suspected Hallerworden-Spatz syndrome – died in hospital (Henri Mondor Hospital, June 1982). It was obvious to the physician in charge that the potential "scientific" interest of an autopsy outweighed the feelings of the family, which was unaware that an autopsy was to take place. Obviously, the knowledge to be gained was of no use to the dead patient, or to anyone, directly. But the findings might possibly lead to a publication, which was the most immediate advantage in sight. Such a case is hardly controversial nowadays. There was a time when autopsies were considered an abuse, and disrespectful of the dead person. There are communities in which this value still prevails. Yet in western societies people who oppose autopsies are considered unenlightened, and there are well-known ways to bypass their opposition. The uncontroversial case can easily be extended to more controversial ones: people who suffer rare syndromes in university hospitals may be investigated so thoroughly, while alive, that they are dispossessed of themselves. Research physicians have all too often become insensitive to human needs in this process.

However, Claude Bernard reasoned as follows: If medicine is to become truly scientific, laboratory experimentation is necessary. Not only anatomical studies on the dead, but physiological studies on live organisms, *i.e*, vivisection, are required. Systematic vivisection for scientific purposes ultimately implies the suffering and death of large numbers of living beings. Common people (including practicing physicians devoid of a scientific mind) are revolted by the idea that living beings should be sacrificed for scientific knowledge. The scientist understands and respects the prejudices of lay people; he does not concern himself with their judgment; he cares only to be judged by his peers. He himself practices a different ethics from the layman. Its principles are: (1) it is "essentially ethical" to experiment on animals (without restrictions) insofar as it is useful for mankind; (2) experimenting on human subjects is compatible with ethics (it is done all the time, not only in medicine); but there is one absolute limit – one should never perform an experiment on human subjects which might be harmful to them to any extent, "even though the result might be highly advantageous to science, *i.e.*, to the health of others" [4].

What has changed since the time of Bernard is that the scientist's choice (of different interests, which imply a different ethics) has become more coercive through society's sponsorship, and scientific progress is presented as if it were an obligation. In the past the choice to uncover new knowledge was an individual choice; as such it did not need to be justified. Whether scientific curiosity was depicted as demoniac and mephistophelean, or as holy, scientific elitism could include a different ethics as long as it was considered an idiosyncratic tendency – to the extent that researchers did no harm to their fellow human beings. Aristotle held that perfect happiness can only be achieved through speculative wisdom and contemplation, an almost divine life, that only very few men care for. A life of rational and political action, if affording only a lesser happiness, is specifically human. In fact, most men identify the good with pleasure and are content with a life of miscellaneous gratifications of the senses [2]. The values of scientific research used to be connected with the leisure, preference, and passion of few individuals. Today the decision to conduct scientific research has become society's decision. Research is largely funded by public money; whether or not the individual citizen approves, a portion of his taxes go to support that research.

Society's choice is not devoid of ambiguity – an ambiguity largely encouraged by professionals. Scientists have often claimed that "pure" or fundamental research is essential, and that it has no other purpose than the advancement of knowledge. Elie L. Wollman, of the Institut Pasteur in Paris, insists that it is a mistake to believe that "science will solve our problems": it creates as many problems as it solves. Pasteur's work has been considered paradigmatic of the ways science can be beneficial to mankind; yet, says Wollman, it has been a primary causal factor in the population explosion and the worldwide disequilibrium with which we are now confronted ([15], p. 410). "Pure science falls outside the realm of ethics", adds Wollman. So, the scientist's viewpoint is often that scientific knowledge is adequately justified by its intrinsic value. Yet, as soon as there is a shortage of funds, professionals hasten to demonstrate that research is advantageous, and proceed to threaten mankind with images of potential disasters of ill-health and underdevelopment should research projects be interrupted ([9], p. 1105). Society's choice, then, is bound to be equivocal. It is remarkable that numerous research projects are liberally funded, which apparently signal no immediate benefit for anyone – including research highly critical of society itself; even there, however, there are more subtle benefits in terms of fame, cultural influence, Nobel prizes, and international standing. After all, research is a luxury primarily accessible in highly developed countries, the way speculative wisdom (according to Aristotle) is accessible to a man who has satisfied all his primary needs – you like to show that you can afford it. On the other hand, it does not take long until more concrete arguments are proferred. The Dangoumau Report of 1982 [7] quite sincerely states that it would be unethical not to seek scientific progress in the field of pharmacology, because (1) one should want to find better ways to alleviate people's suffering, and because (2) scientific progress is synonymous with economic progress, including the development of industry and increased employment.

It is doubtful whether society at large can adhere to what Jacques Monod [16] called "the ethics of knowledge", the central principle of which is that the search for objective knowledge should be promoted. The ethics of knowledge is puritanical; its "fierce and austere" rule justifies sacrifice and suffering. It carries the idea that we were not born to enjoy ourselves but to produce important works that transcend us. Scientific progress is indefinite, there is no end to it, and it necessarily includes emotional asceticism. It is remarkable that such an ethic could be even partially assimilated within society. The recent trend in educational reform is that society should generalize the scientific methodology and viewpoint. The U.S. National Commission for the Protection of Human Subjects of Biomedical and Behavioral Research insisted that it did not adhere to a utilitarian value of science. Its stated perspective was that in some cases, indeed, research projects aim explicitly at solving medical or social problems. This usually implies a careful analysis of what kind of benefit is anticipated; for whom, at what cost and risk; with what chances to succeed, and what is needed to evaluate the project. In other cases, however, research protocols are developed to yield theoretical knowledge. Then it is pointless to speculate on its potential use. Evaluation of the project will usually include a careful review of the methodology, and – given the consent of subjects of the experiment – take account of the consequences only if they may seriously endanger public health or safety ([15], p. 225).

The ambiguity, however, can't be eliminated. Researchers take it as obvious that scientific progress is thrilling and valuable, nay inevitable, fulfilling man's evolutive destiny. But even laymen know that doing research is an arbitrary choice, that such a choice has been entirely foreign to large populations. Some of our contemporaries in highly developed countries have argued that such a choice can be extremely costly and of doubtful benefit, even some physicians believe that in some cases certain diagnoses should not be sought.

Paradigmatically ambiguous, for that matter, is the U.S. Belmont Report [17]. The *principle of justice* – which stipulates that the risks and advantages of research should be equally distributed, and, in particular, that the burden of being a research subject should not fall entirely on economically or physically disadvantaged people – implies that some research should be done, and that scientists should manage to formulate an ethic compatible with the consequences of such a choice. Thus it has been suggested that people be drafted to serve as subjects of scientific research. The *principle of beneficence*, however, conflicts to some extent with the latter, to the extent that research protocols are defined as protocols that aim at confirming an hypothesis and elaborating generalizable knowledge. The crucial question is whether the cost/benefit analysis implied by the principle of beneficence is marginal and external to scientific preoccupations proper, or a calculus of the mathematical-expectation type is an essential part of scientific methodology, which ought to be incorporated into research protocols.

3. THE PRESUPPOSITION THAT KNOWLEDGE (APPLIED TO HUMAN BEINGS) IS INNOCENT

The second presupposition often involved in research projects is the presupposition that knowledge *per se* is innocent, or even curative and intrinsically beneficial, *e.g.*, when freeing us from prejudice.

Researchers indeed know that the methods by which knowledge can be acquired are by no means innocuous, and when applied to human beings the range of possibilities is severely limited by ethical considerations. Various committees have observed that the risk involved in being a subject in an experiment is, in fact, very low and compares favorably with occupational risks. But that is because there has always been spontaneous regulation on the part of the scientific community, whereas current regulations have yielded only minor changes.

Of course researchers also know that the application of scientific knowledge is apt to be devastating (*e.g.*, military applications).

But scientists usually assume that knowing *per se*, as well as divulging the knowledge – for scientific knowledge is public – can do no harm either to the knower or to the object known, provided that some confidentiality is assured. *Homo sum et nihil humani a me alienum puto*, Terence said. As a consequence, when we come to the study of human beings, there are no taboo topics – scientific methods of analysis can be applied to virtually anything. It amounts to assuming that scientific knowledge is objective, neutral with respect to common values, unbiased because it is open to criticism and verification, and that interactions between observers and the observed can in no way be prejudicial to either party.

Such a perspective has been fervently challenged; *e.g.*, some research on I.Q. or XYY chromosomes has been judged damaging for the subjects under study and discontinued. Researchers have even been persecuted in the name of ethics for pursuing such knowledge. But most scientists tend to think that these were ideological issues, and that research would soon be resumed when the situation calmed down.

However, we also know that the image of "objective" knowledge is largely an idealization, especially when it comes to the knowledge of human beings. Important scientific movements, for example in comprehensive sociology, have been based on the assumption that knowledge can't be purely objective and neutral.

In fact, it is well known that (1) scientific knowledge cannot be easily dissociated from the means by which it is acquired: the content varies with the methods of acquisition, which is especially obvious in "clinical" psychology *vs*. "experimental" psychology. And (2) scientific knowledge cannot be easily dissociated from the consequences: psychoanalysis is an extreme example of therapeutic acquisition of knowledge, but medical doctors often speculate on the influence that a patient's level of knowledge about his illness has on the course of illness, for example.

There is an inconsistency in holding that knowledge is valuable in itself and that it is neutral with respect to values. One certainly does not want to go as far as Monod's saying that "science is destructive of our values", but there are numerous examples of the ways in which knowledge can insensibly modify common judgments and their reference values ([10], §§ 4, 5). One example is how the discovery of the "mechanism" of the "pre-menstrual syndrome" in women, and the efficacy of specific drugs on the syndrome, has undermined the traditional contempt for feminine periodic nervousness.

To the extent that researchers recognize that the acquisition of knowledge about human beings is not entirely neutral and innocent, they have to admit that there are circumstances in which acquiring knowledge may be untimely or pernicious in itself. Given the current limited state of therapeutics, for example, a research aiming at allowing doctors to diagnose multiple sclerosis at an earlier, preclinical stage might not only be useless but devastating to the patient[1]. One could argue that such knowledge would not be harmful if it were not used, but it is incoherent to claim that one may acquire such knowledge and not use it (if only because in the process of acquiring it, some subjects would be diagnosed as potentially having multiple sclerosis). A similar argument was developed [5] about the possibility, raised by some investigators, to elicit premature symptoms of Huntington's chorea for diagnostic purposes, or even to conduct mass screening programs aimed at identifying presymptomatic carriers of the gene. The rationale would have been that Huntington's chorea is a genetic disease that becomes symptomatic only in adult life, usually after the carrier has passed on the morbid risk to his progeny. But the social benefit did not seem to compensate for some individuals being faced with a prognosis of dementia.

4. THE PRESUPPOSITION THAT THERE ARE PROBLEMS OF A HIGHER OR LESSER INTEREST

It would be dishonest if not circular to argue that a problem like diagnosing and informing patients of multiple sclerosis or Huntington's disease at a subclinical stage has "little scientific interest anyhow", and that "there are other more urgent scientific priorities", as though real scientific "priorities" always coincided with some good – which is precisely what is being questioned.

The third presupposition is, in fact, that some problems have, at some point, a greater or lesser scientific interest or urgency.

Evidently some choices have to be made about the appropriateness of engaging in certain research projects. How are scientific priorities determined? Cynics answer: look and watch where the money goes. First comes research connected with the military. Second, far behind, is medical and/or industrial research. Last is the miscellaneous educational research, *e.g.*, psychology and the social sciences. But even if this is the case, who is supposed to make the decisions of priority?

The apparent issue here is the conflict between democracy and elitism. It was a pressing issue some years ago. Scientists like to emphasize freedom of investigation, and to stress that only specialists are apt to appreciate the scientific importance of various projects. Otherwise, they argue, we will see the tyranny of incompetence. On the other hand, in the early seventies there was a powerful "anti-science" movement to deny scientists the right to determine what kind of research is appropriate. Radical thinkers wanted "science for the people" and claimed that "there are more urgent scientific problems than the existence of quarks". Conservatives would say that research is usually quite costly, counterproductive, and a source of social turmoil. Of course, a negative control was exerted – cut the funds, which is particularly effective during periods of recession.

A more relaxed atmosphere seems to have prevailed recently. It appears that society will permit the scientists' viewpoint to prevail. In France, in December 1981, the Ministry of Research organized a broadly-based national consultation of researchers under the *Colloque Recherche et Technologie*, whose aim was to define important research objectives and priorities that would reconcile democracy with professional jurisdiction. The U.S. National Commission for the Protection of Human Subjects of Biomedical and Behavioral Research insisted that particular research protocols could be evaluated according to whether or not their methodology was sound, and guaranteed that a conclusion could be obtained which would significantly contribute to our knowledge – irrespective of whether potential benefits could be expected for anyone. Saying that research on the fetus *in utero* can be conducted if "the knowledge to be gained is important" [17] is tantamount to permitting researchers to use their judgment and to pursue what they consider important.

After a period of "cultural revolution" during which researchers were reminded that they could not both obtain funding by the community *and* do what they wanted, apparently things returned to where they had been, *i.e.*, the competence of professionals is recognized. It is as though society were saying: "It is essential that you pursue good science, and you are the ones who know what that means; we shall merely make certain that you will not seriously harm the human subjects we allow you to use, and that your methodology is proper, *i.e.*, complies with your own standards" – for scientific standards are those internal to the scientific community itself.

The assumption that there are objective criteria with regard to what counts as good scientific methodology is of course debatable. There may be clear negative criteria for what is unscientific, but the existence of positive criteria for what is to count as true is more doubtful, as was argued by Popper [18]. However, the fact that the assumption is "ratified" by political or administrative instances seems to suggest that researchers have indeed emerged victorious. But while researchers feel relieved and content with the power they have gained, they may well be manipulated by political and economic forces far beyond their control. However, this intuition should be balanced against the conflicting intuition that scientific development is a natural process that can be forced artificially only within limits, like the growth of animals and plants.

5. THE PRESUPPOSITION THAT HUMAN BEINGS SHOULD BE USED SPARINGLY IN HUMAN RESEARCH

The fourth presupposition is that human subjects should be used sparingly in research, *i.e.*, used only in experiments involving minimal risk, or a risk which compares with occupational hazards, or the hazards of everyday life, and following on a careful review of the relevant literature and truly sufficient laboratory and animal experimentation (this applies especially to biomedical research, where the major risks are supported by animals).

Note that there is an (admissible, as long as it is conscious) arbitrariness in the evaluation of what counts as "minimal" risk and "sufficient" preliminary experimentation. In pharmacological trials, for example, it is currently assumed that the potentially teratogenic effects of a new drug should be studied on three different animal species before (Phase 1–2) preclinical trials on human subjects can be tackled. Why *three*? It is an induction. So far, teratogenic substances in humans have been found teratogenic in at least one species out of three studied: so, three seems like a safe bet. Note also that insofar as the human costs consented to are taken into account, "truth" and scientific progress are valued less than some other human ends. "Truth isn't worth such a price," some say. The price of truth is less than the price of improving the quality of automobiles (at the cost of automobile competition) or the price of defending national interests and pride (at the cost of war).

The use of animal models calls for a few comments.

It is difficult both to use animal resources for human research purposes and to deny the fact that scientific research is more anthropocentric than purely objective. On the other hand, it has often been said, à propos of human experimentation, that the end does not justify the means ([3], p. 1360). But clearly in the case of animal experimentation it is presupposed that the end does justify the means. This signals yet another ambiguity or incoherence.

The rationale by which we feel justified in using animals in our scientific endeavors is (I think) the same rationale by which, in a not very distant past, researchers felt justified to experiment on the poor, the uneducated, the old, the mentally or physically ill, foreign and/or "underdeveloped" populations. Sociobiologists have interpreted such behavior as a form of genetic selfishness, the reverse side of the advertised beneficence of science. Benefits are sought, they argued, at the expense of others, for individuals of the same breed. In fact, it is currently admitted that animals should be conserved as *species*, *i.e.*, one should strive to conserve samples of all varieties in their ecological niches. At the same time it is assumed that *individual* animals are valueless, and that we may draw samples from the stock whenever there is an abundant supply, and manipulate their breeding for laboratory purposes. In this case laboratory work seems as important as mere survival, which justifies breeding and killing animals. By the same token one might say that scientific research has had a tendency both to conserve samples of human cultures (*e.g.*, for ethnologic studies) or human pathologies (think of the "natural experiment" in Tuskegee, or of "historical" cases in psychiatry who go untreated for a while, so that students may see what a "typical" mania or hysteria looks like), *and* to waste human stock devalued by its relative unproductiveness and abundance.

This leads to a "paradox" ([15], p. 359). A "good" animal model must be assumed sufficiently "close" to the human breed to authorize inferences from animal to man, and sufficiently "distant" to allow guilt to remain unconscious or discharged. Thus the "genetic distance" is not measured on the same scale for the purpose of science or ethics. The "other" is assumed to be far more different ethically than it is scientifically.

Should such a presupposition apply only to animals, one might argue that it is a choice about which there is a *de facto* consensus, a consensus large enough so research is not significantly threatened by sporadic minority protests which identify animal experimentation with torture. But this presupposition applies to human subjects as well.

For instance, a study was published recently on stress-induced analgesia. The study was performed by a team of competent clinical neurophysiologists. The experiment was sophisticated in design, astute, and demonstrative. (Publication in a scientific journal more or less guarantees that the protocol was scientifically and ethically admissible, and that the statistical treatment was correct.) The subjects were normal adults, fully informed, and had given their consent in writing. They were administered electric shocks on the lower limb, inducing "severe pain" ("7 to 8 times the threshold of pain"). They were told to memorize the "intolerable pain" as a reference. Then the effect of anticipating another such shock, *i.e.*, being in a state of stress, was investigated. A progressive rise of the threshold was observed as the stress was repeated. This effect was entirely reversed by an injection of naloxone (and not by the injection of a *placebo*) ([24], p. 1389). In an other than "scientific" setting such an experiment would be considered torture. Indeed, the subjects had given their consent. Note that the experiment was performed and published in a country where experiments on healthy volunteers were not legal at the time, and jurisprudence tended to consider such consent of the subjects null and void. A consent to what, anyhow? They could not know what their reactions would be, for that is precisely what was being investigated. They could not control them and, of course, they were expected *not* to control them. Being a sample, the reactions of which are dissected, is worse than being exploited, remarked Hans Jonas. The researchers comment coldly that "the injection of naloxone induced some behavioral side-effects, such as aggressive reactions (in three subjects) and weeping (in one woman); these effects lasted from 15 to 20 minutes". In such experiments, the human reactions are turned into effects. The subject is a substratum. He can only feel stupid, or vexed, *unless* he identifies himself fully with his "objective" observers, and puts a distance between himself and the observed self, a point excellently made by Hans Jonas ("principle of identification":[12]).

In many scientific investigations using human beings, everything is done to "neutralize" the effects of a subject's aptitude to act as a fully competent person. Some information is withdrawn, or the subject is prevented from exercising judgment, initiative, critical powers, fantasy, intelligence. It is doubtful whether this is always a "scientific" necessity, *i.e.*, whether this is done only to reduce variability. It may be a sort of "ethical" necessity as well, destined to create a distance between the experimenter and the subject – a distance the researcher may need whenever he would not want the roles between the subject and himself reversed. The subject may believe that the experiment is beneficial for himself. Note that making him believe so wrongly is a way of inducing the distance – whoever entertains what I know to be false beliefs I tend to consider inferior. Unless the subject believes that he will benefit from the experiment, the only way for him to restore his dignity as a person, as Jonas has pointed out, is to become a spectator of himself, at the same time accepting to be that part of nature which is being investigated. This requires wisdom, in the Schopenhauerian sense. Psychoanalysis, insofar as it claims to be a scientific practice, also claims to induce such an attitude.

As research develops, we may come to judge that a few experiments on intelligent and cooperating subjects can teach us as much as a great many experiments on (induced) imbeciles. But this would call for a greater methodological imagination, statistical treatment of small samples, rigorous analysis of trials which are not blind, etc. Some such procedures have been suggested by Lindley [13].

6. THE PRESUPPOSITION THAT IT IS UNETHICAL TO CARRY OUT SUBSTANDARD RESEARCH (ERRING IS UNETHICAL)

The fifth presupposition is that error is manifold and there is only one way to truth. One should strive to eliminate errors, and choose the best research strategy. Let us first look at what is involved in eliminating errors.

It is relatively clear in general what counts as an *error* in methodology, and usually easy to recognize that it is *unethical* to carry out poor scientific experiments, particularly on human beings. In poor experiments (1) subjects are misused, and researchers' time and energy is wasted, which goes against the principle of respect for persons. And (2) if ever the results of such experiments are published they may induce others into error. The consequences may be devastating, either because other research teams will waste time and energy trying to reproduce the results, or because the publication will remain unchallenged, influence practical attitudes, and it may even be considered unethical to carry out further trials – thus possibly denying some patients the right to obtain the "better" care [1].

Yet Douglas Altman observes that researchers pay the principle "mere lip service". In fact, as has been shown by several authors (*e.g.*,[23]), a large proportion of published research is substandard, and the mishandling of statistical methods affects up to 50% of published papers. Most errors would be easily detected and remedied, should researchers have a firmer background or the opportunity to cooperate with professional statisticians, and should IRB's and professional journals raise their standards.

Altman proceeds to provide examples of errors frequently committed by biomedical researchers at the various stages of research:

(a) Design of the Study

When the characteristics of subjects eligible for the study are not precisely defined, extrapolation of the result to the population of interest is doubtful. When the study design is not completed by a pilot study, major difficulties are not anticipated and the project may have to be aborted when encountering such a difficulty. "No amount of clever analysis later will be able to compensate for major design flaws", Altman writes ([1], p. 2). And May remarked: "a poorly designed or poorly conceived experiment is unethical by definition and should not be permitted" [14].

(b) Sample Size

There is a definite relation between the power of a statistical test (*e.g.*, the chances to detect a certain effect), the size of the effect to be detected, and the sample size. Apparently few medical doctors understand the meaning of statistical notions: inadequate samples for clinically interesting effects to be detected are commonplace. "How far (is one) prepared to alter one's criteria of acceptability for the sake of expediency?" Altman asks ([1], p. 3).

(c) Collecting and Screening the Data

Too often subjects who drop out of a study are discounted, whereas they in fact constitute an atypical subgroup; or "outliers" are eliminated without further examination; or only "interesting-looking" cases have been examined. A certain degree of sophistication would rule out such biases which prematurely distort meaningful data.

(d) Analysis of the Data

Many researchers tend to ignore that most standard statistical tests rest on the assumption that the data are *normally* distributed, or are confused about the difference between correlation and regression analysis – a confusion possibly exacerbated by the use of computers.

(e) Interpreting the Results

It is a mistake to believe that one has *demonstrated* that there is a *significant* difference, when the difference is found significant (say) at the .05 level. The level of significance .05 is not the probability that the difference is real; rather, it is the probability of finding a difference *if the null hypothesis* (of no actual difference) is true – but the probability of the null hypothesis is generally unknown. The decision to assign the observed difference in the results to the difference in treatments is *always arbitrary to some extent*, even if all possible variables have been carefully controlled. Physicians are prompt to jump to conclusions because they employ all sorts of background information and indices external to the study itself; this is permissible, provided they are aware of the extent to which statistical models fail to support the conclusion.

Fortunately, consultation with professional statisticians is becoming more frequent. But the dialogue between clinicians, whose first preoccupation is presumably the welfare of their patients, and laboratory researchers, who are interested in "fundamental" problems, is already difficult. The intervention of statisticians, who tend to introduce complicated mathematical requisites little understood by others, makes it even more difficult.

One is certainly entitled to deplore the substandard quality of much methodology reflected in scientific papers, and to ask on the basis of ethical considerations that it be remedied. However, (1) the reluctance or negligence of experimenters to comply with certain "official" standards may be a sign that the standards are inappropriate – or poorly understood, *i.e.*, assumed more rigid than they are. Altman himself points out that researchers should know when *not* to use a significance test, and when to refrain from drawing a meaningless line across a scatter plot of data. When comparing the effects of tricyclic antidepressant drugs *vs.* psychotherapy on depressive patients, why systematically add a placebo drug to the psychotherapy, Hollon and de Rubeis ask [11]. As Altman remarks, "there is no one best design for *all* clinical trials" ([1], p. 2). Moreover, (2) perfect statistics and a rigorous experimental procedure do not imply that you have the concepts right. Orthodox statistical procedures are designed to test hypotheses: they are no guarantee of the pertinence of hypotheses. It is certainly true that a cleverly drawn hypothesis deserves a well-designed and competently carried-out experiment, but good experiments consume resources and human energy and would sometimes deserve that the basic ideas be more original than they are. (I have often thought that certain experiments, *e.g.*, in psychology – learning experiments, for instance – are an insult to both the subjects experimented on and mankind in general, due not to a sloppy methodology, but to the simplistic underlying assumptions. No methodological rules can protect us from stupidity.)

7. PRESUPPOSITIONS INVOLVED IN THE CHOICE OF THE "BEST" RESEARCH STRATEGY: THE ORTHODOX "EXPLANATORY" APPROACH vs. THE "PRAGMATIC" APPROACH

Devising the best research strategy requires that objectives be made explicit. The sixth presupposition is that the choice of the correct scientific procedure can be standardized. This presupposition is perfectly acceptable, provided the finality of research is taken into account. (Technical imperatives are merely what Kant would have called hypothetical imperatives.) It is important to realize that different ends-in-view may imply different scientific methodologies.

Consider the example of Phase 3 clinical trials. Typically, a new treatment B (or diagnostic test) is compared to a reference treatment (or test) A, such as a codified standard treatment, or placebo, or even the absence of treatment. What exactly does one want to know? Pragmatic medical minds are interested in knowing whether they should switch to treatment B, or continue to prescribe A to patients with disease Z. Theoretically-oriented researchers are interested in knowing whether treatment B is more effective than A on disease Z itself. Daniel Schwartz et al. [20] have emphasized the fact that orthodox statistical methodology implies an "explanatory" or "research-oriented" viewpoint, whereas many if not most clinical trials involve the pragmatic motivation of providing practicing clinicians with a basis for decision concerning the choice of therapy, which may be better served by statistical protocols different from the ones routinely used. Thus, to devise an appropriate experimental strategy to solve a given problem involves a reassessment of orthodox and alternative procedures[2]. Let us now sketch a research-oriented strategy and a pragmatic strategy with their methodological implications.

(a) Research-Oriented Strategy

The theoretically-oriented researcher wants to know whether or not "there is a significant difference" between the effects of A and B. The hypothesis tested may be, for example, the null hypothesis: there is no difference.

In this perspective, the worst scientific outcome would be to conclude that there is a difference if actually there is none (type I error – to select a false conjecture). Set $\alpha = 0.05$ (for instance) the maximum probability α you accept to commit type I error. Given small α, you then want to minimize the probability β of not detecting a real difference (type II error – to reject a true conjecture). This can be achieved with a sufficiently large number of subjects. The number n of subjects necessary for the trial to yield a conclusion (sample size) is inferred from the smallest difference Δ between the effects of A and B such that you do not expect to detect true differences less than Δ within the specified error rates. When α and β are small, then γ – the probability to commit type III error, *i.e.*, to mistake the worse treatment for the better – is negligible, and can be ignored safely.

This orthodox statistical approach is particularly appropriate, for example, to resolve a causal question, such as: "does the administration of B modify the course of disease Z?" or "how active is B on disease Z?"

The experimental setting will have to efficiently serve the purpose of answering such a question. The theoretically-oriented physician prefers to test B against the absence of treatment, or against a placebo. Subjects are the material from whom signs of actual activity can best be extracted – the "subject-as-a-means". Subjects eligible for the trial must be as homogeneous a set of patients as possible, who exhibit a typical form of the disease, and who are presumably sensitive to treatment. It is important to reduce the number of possible withdrawals of subjects, for such withdrawals would upset the statistical analysis. Hence, the criteria of selection are rigorous, and the patients are carefully observed. The treatment is administered under rigid laboratory conditions, preferably in hospital. The conditions of administration tend to optimize the "biological effect" of the substance, as assessed in the pre-clinical (phase 1 and 2) trials (*e.g.*, intravenous injections every 4 hours), so as to give the drug or chemical substance the best chance to "prove" its effectiveness. The approach is analytic: it attempts to isolate the desired effect, and to discount adverse or side effects unless they are serious. Observed differences will be attributed to the treatment only if the test and control groups are strictly comparable in all respects except for treatment. Systematic randomization after the selection of eligible patients guarantees initial comparability, but comparability must be maintained throughout the experiment by methods such as blind or double-blind administration of drug, neutralization of placebo effect, blind assessment of results. For assessing the outcome one chooses one or a small number of "objective" indices with "biological" meaning if possible (*e.g.*, histology on biopsy sample), or at most only one complex criterion (multivariate analysis). The method for comparing the two groups is a standard statistical test of hypothesis.

A *theoretical conclusion* will be reached in any case, whether one achieves a positive or negative result. It involves (1) a "judgment of significance", based on the statistical test, *e.g.*, there is a probability less than .05 that the observed difference is the result of mere sampling variability; and (2) a "judgment of causation" that the observed difference is attributable to treatment B. The theoretical conclusion yields a clear *practical conclusion* only if the result is negative, *i.e.*, the difference is "not significant". A positive result – the difference is "significant at the . . . level" – does not imply that treatment B should be preferred by practitioners! So, the net practical gain is to *invalidate or falsify* some hypotheses, *e.g.*, demonstrate that laetrile is *not* an effective treatment of cancer, and such a negative conclusion has a claim to universality.

There is a cost-benefit analysis underlying the "explanatory" attitude. *The cost* (implicit in the choice of α, β, Δ) of reaching a wrong conclusion (and publishing it!), and/or missing an interesting finding, with an evaluation of what counts as an interesting finding, *is weighted against the cost* (implicit in the choice of n) of experiment's time and resource spent. Ethical rules concerning the respect due to subjects function as external constraints. There are ethical impossibilities just as there are physical impossibilities. This calls for strict ethical rules that fix the *limits* to what may or may not be done. Within the limits, however, researchers do not have to bother with ethical considerations.

(b) Pragmatic Strategy

The pragmatically-oriented physician wants to know which of A and B is the better treatment. If in actuality A and B are equivalent, it does not matter for the patient whether he selects A or B. Thus, type I errors are of no practical importance, and there is no point in minimizing α. Set $\alpha = 1$. Any difference is significant at that level. A significance test is useless. "We simply choose whichever treatment gives the average better observed result." Note that "$\alpha = 1$" means "we are always in error if A = B." That is sensible from a practical point of view. If we must choose between two strictly equivalent solutions, and if choosing amounts to saying that the chosen solution is better, then choosing is erring whenever the solutions are equivalent. Another way of looking at this is:

we must reach a conclusion at the end of the trial. Failing to conclude in favor of the new treatment amounts to concluding in favor of the old one – the conservative conclusion. Therefore we want $\beta = 0$ (in fact, $\alpha = 1$ if, and only if, $\beta = 0$). Now the error rate of the third kind is by no means negligible. From a practical point of view it would be a serious error to recommend the worst treatment. The number of subjects n is chosen so as to minimize γ.

The experimental setting will have to serve the purpose of answering the question: "Is the new treatment an improvement on treatments currently available?" One prefers to test B against the best current therapy. The end-in-view is to maximize the benefits for the patients – the "subject-as-an-end", although an end as a class rather than as an individual. The subjects eligible for the trial are a representative sample of the population to whom the results are to be extrapolated. It may be a more heterogeneous set than in the previous case; the criteria of selection may be more flexible (*e.g.*, subjects with two pathologies not excluded). One should avoid trying within the hospital a drug meant to be used by patients at home, and those subjects who drop out are one aspect of the outcome. The selection of subjects prior to randomization and the process of randomization are features common to both approaches. Here, however, it may not be possible, or even desirable, to maintain comparability through blind procedures, or perhaps the blind methodology may be adjusted, for instance along the lines described by Zelen [27]. Treatment is administered under conditions realizable in practice; policies are flexible (*e.g.*, doses may be adjusted) to give patients the best chance to benefit from therapy and to optimize the "practical" effects. This approach is synthetic or global: the context (*e.g.*, side or placebo effect) is "absorbed" and the conditions are optimized in either group. For assessing the outcome, the criteria of choice are those which are of direct interest to the patient, *e.g.*, either one overriding and inclusive criterion like the duration of survival or return to work, or a weighted combination of criteria like an inclusive summary possibly describing the quality of survival. Rather than the effect of treatment proper, what is evaluated is the course of the treated disease. So-called "side-effects" are thus integrated into the effect. No statistical test is necessary. The relevant schemes of reasoning belong to decision theory.

A *practical conclusion* will be reached in any case: either B proves definitely better than A, and one should switch to B, or B does not prove definitely better, and one remains with A. A *theoretical conclusion* is attained *only* if the result is positive, *i.e.*, if B is better than A. The fact that B does not prove better than A in practice does not imply that it is not more active than A, for it may be more active, but have greater side-effects. The new knowledge is more synthetic and less specific than before. The discovered "truth" has no claim to universality, for the findings might be different at different times and/or places, and results may safely be applied only within the population of reference.

There is a cost/benefit analysis underlying the pragmatic attitude. *The cost* (implicit in the choice of γ) of an error of sign, *i.e.*, recommending the worse treatment, which depends on the probability and seriousness of ill consequences of worse treatment, and on the size of the population of reference to which the recommendation will be applied, *is weighted against the cost* (implicit in the choice of n) of having $2n$ patients in a trial, n of whom are receiving the worse treatment, which depends on the duration of the trial and the size of the group missing the better treatment. The optimum number of subjects is the best compromise between requirements of the first kind, which call for a large n, and requirements of the second kind, which call for a small n. "Both kinds of cost are essentially of an ethical nature and can be expressed in terms of the number of patients suffering potential harm" ([20], p. 118). Sequential designs will be used whenever possible, *i.e.*, when the outcomes can be assessed relatively quickly: sequential designs imply that the larger the difference between the effects of treatments, the smaller the sample size will be, hence the fewer subjects receive the less successful treatment, which minimizes the cost of the second kind. At any rate, ethical considerations are integrated *in* the scientific procedure, which is viewed as a decision process. It is desirable that the procedure be explicitly analyzed as a decision process, *i.e.*, that the utility and probability functions be evaluated and the mathematical expectations computed. To the extent that such evaluations are arbitrary and/or cannot be made entirely explicit, researchers are expected to use their judgment thoughout the procedure; for no "strict" ethical rule can solve all problems in advance. This calls for an ethics of responsibility, wherein one is aware of the fact that ethical principles are contingent on the choices made of what we want the "good" of humanity to be.

The differences between the two preceding approaches are clear. In the first, we wish to know from a biological point of view whether a difference exists or not: to conclude that one exists when this is not so is an error. Rather than make such an error, we may sometimes be prepared to reach no conclusion. The solution is then provided by a *significance test*. In the second approach, we require to choose one or other of the treatments. To choose one when it is equivalent to the other is perfectly acceptable. It is essential to reach a conclusion and we do so without performing a test: we are involved with a decision problem ([20], p. 83).

Of course, the distinction drawn is schematic. One might say that the two attitudes described above broadly correspond to two types of problems. It would be erroneous to conclude that one is more "scientific" than the other. In sum, "purely" theoretical problems, *e.g.*, to determine whether a slight change in the molecular structure of a substance modifies its biological activity, are intended to advance "fundamental" knowledge, whereas "purely" practical problems, *e.g.*, to compare two drugs of two different chemical families, or two protocols of treatment, to find out which one is more beneficial to patients, are intended to advance clinical knowledge. However, most problems in biomedical sciences are of an *intermediate* type – both theoretical and practical. Ideally the "correct" strategy should thus be a mixture of the two "pure" strategies. That is hardly possible rigorously, although it may be expedient. An example given by Schwartz *et al.* is this: In a "pragmatic" trial on the prevention of infarcts through diet, since the relevant criterion – incidence of infarcts – involves long delays to be correctly assessed, one may wish to replace it with another criterion known to be associated with it, *e.g.*, a drop in the blood cholesterol level, but one should be aware that it implies switching to a more "explanatory" and less pragmatic model. Strictly speaking, in order to obtain a result for each approach, two trials with different designs will be needed. However, it will often not be realistic to do both; either there are not enough subjects available, or the chances are too high that yet another treatment will be discovered that overrides both A and B before the trial is completed. Therefore, one must choose in each case which question one wants answered. "Should we aim to obtain an immediate increase in knowledge, hoping for eventual application, or should we aim at immediate practicality which is based upon incomplete understanding and so perhaps is less promising for the future?" ([20], p. 33).

The "correct" scientific strategy is the strategy appropriate to the end-in-view, the one apt to give a clear answer to the stated problem. The important point is to see clearly the objectives, constraints, and limits of these strategies.

8. PRESUPPOSITIONS INVOLVED IN EXTRAPOLATION OR GENERALIZATION OF RESEARCH

The seventh presupposition bears on the generalization or extrapolation of results. Being able to generalize is an ambition of science. At the same time extrapolation is implicitly or explicitly normative, and leads to what may be called the "tyranny of rationality": treatment B becomes the standard treatment, *e.g.*, doctors prescribing other treatments may be blamed, or the population at large is vaccinated, or people are persuaded to switch from butter to oil, etc. Extrapolation is easier (immediate, but perhaps more cautious, for limited to population of reference) in the pragmatic approach than in the "purely" theoretical approach (where the only legitimate extrapolation is of "negative" results – but then it tends to be radical). As researchers are called on to play the role of experts, they can hardly claim that extrapolation of positive results, which leads to applications, is entirely the politicians' responsibility, since the scientists deliver the relevant information. It seems that researchers are actually prompt to exert the power offered to them, devising the rules of a "rational" lifestyle ("You should exercise; You should not smoke; You should not eat salt or animal fat"), and sometimes imposing very severe disciplines (*e.g.*, lifelong treatment of mild hypertension), even through state regulations (safety-belts) and health policies (mandatory vaccination). However, when unanticipated problems arise, researchers seem equally prompt to withdraw from the pragmatic attitude, to recall that they had *only* statistical results, and to transfer the responsibility to others.

The normative aspect of extrapolation is based on the presupposition that everybody wants to make rational choices and maximize expected utility. This is a massive postulate of health policies as well as scientific medicine. Of course one hears that, *e.g.*, when sick, one need not seek medical care ("You are perfectly free to consult a quack rather than the best specialist"). But the way the question is framed amounts to leaving people with the alternative of complying with the rules of "rational" life or being abandoned to the perils of ill-health (or even penalized: if you get lung cancer and are a cigarette-smoker you should feel that it is your fault, and moreover, it justifies that you paid a high tax on each box of cigarettes you bought, to compensate for what your illness costs to society). The nuance is perhaps that "irrational" behavior and self-inflicted morbidity are presumed perverse by those with pragmatic tendency who may bend towards an ethics of "victim-blaming", while others ("pure" scientists?) tend merely to consider it stupid and take refuge in an ethics *à la Pilatus* ([8], p. 26 *sqq*). Of course, the limits to a "scientifization" of life can also be recognized within scientific practice. The Dangoumau Report says of traditional therapies, the value of which cannot easily be assessed by current scientific techniques, that it would be unreasonable ("unethical") – even though, perhaps, rational – to exclude them from medical prescriptions "because our contemporaries don't nourish themselves of rationality only" ([7], p. 17).

It is known that when administered to a large population in phase 4 trials, a treatment (or any systematic procedure) will have effects that could not be anticipated – long-term effects, rare or late side-effects – because of the necessarily limited character of scientific trials, and the uncertainties of induction about natural kinds within which variability is large. At the time of extrapolation, it is assumed that the late or rare (unknown) detrimental effects will be less serious than the withdrawal of treatment. This assumption is maintained as long as the treatment is in use. This means that even physicians presume that an individual's life has a price, that its value is not infinite, and that the majority's welfare justifies that a few persons be sacrificed – a view often attributed to "pure" researchers. In other words, it means that not only the benefits and disadvantages *for* persons, but also the value *of* persons' lives (either the quantity of life or its quality), is included in cost/benefit analyses of human research, just as it is in other human activities. That is one reason to believe that a "minor" risk is interpreted not only as any probability of mild inconvenience, but also as a low probability of disastrous effects. The idea that people's lives have a limited value is very foreign to physicians' traditional explicit, and perhaps irrationally naturalistic ethics, one basic maxim of which has been *primum non nocere*, and not "seek progress, even at some cost to persons."

ACKNOWLEDGEMENTS

The author is very grateful to Stuart Spicker for his competent and cordial editorial assistance.

Université de Paris, X
Nanterre, France.

NOTES

[1] The multiple sclerosis example is due to Dr. J. Poirier, Hôpital Henri Mondor, 94000 Créteil, France.

[2] In part 6 I draw heavily from [20], with the kind permission of the authors, translator, and publishers.

BIBLIOGRAPHY

1. Altman, D. G.: 1980–1981, 'Statistics and Ethics in Medical Research: A Series of Eight Articles', *British Medical Journal*, 1980: **281**, (1) 1182–1184, (2) 1267–1269, (3) 1336–1338, (4) 1339–1401, (5) 1473–1475, (6) 1542–1544, (7) 1612–1614; 1981: **282**, (8) 44–47.
2. Aristotle: *Nicomachean Ethics*, Book X. Ed. Bekker, 1831, II, 1172–1181.
3. Beecher, H. K.: 1966, 'Ethics and Clinical Research', *New England Journal of Medicine* **274**, 1354–1360.
4. Bernard, C.: 1865, *Introduction à l'étude de la médecine expérimentale*, Paris; English translation by H. C. Green (1957), *An Introduction to the Study of Experimental Medicine*, Dover Publications, New York.
5. Brackenridge, C. J.: 1981, 'Ethical Aspects of Plans to Combat Huntington's Disease', *Journal of Medical Ethics* **7**, 24–27.
6. Code de déontologie médicale, 30 Juin 1979, Décret n° 79–506 du 28 juin 1979, *Journal Officiel*, Paris.
7. Dangoumau, J.: 1982, *Expérimentation clinique: essais thérapeutiques, pharmacovigilance, pharmacologie clinique*. Rapport rédigé pour le Ministre de la Santé, Paris.
8. Dworkin, G.: 1981, 'Taking Risks Assessing Responsibility', *Hastings Center Report* **11** (5), 26–31.
9. Eisenberg, L.: 1977, '*The Social Imperatives of Medical Research*', *Science* **198**, 1105–1110.
10. Fagot, A.: 1982, 'Science et éthique', in G. Fløistad (ed.) *Contemporary Philosophy: A New Survey*, Vol. 3, pp. 277–349, Martinus Nijhoff Publishers, Haag, Holland. Reprinted in Fagot-Largeault, A. (1985), *L'homme bio éthique*, Maloine, Paris.
11. Hollon, S. D. and DeRubeis, R. J.: 1981, 'Placebo-Psychotherapy Combinations: Inappropriate Representations of Psychotherapy in Drug-Psychotherapy Comparative Trials', *Psychological Bulletin* **90**, 467–477.
12. Jonas, H.: 1969, 'Philosophical Reflections on Experimenting with Human Subjects', *Daedalus* **98**, 219–247.
13. Lindley, D. V.: 1965, 'The Effect of Ethical Design Considerations on Statistical Analysis', *Journal of the Royal Statistical Society*, C 24, **2**, 218–228.
14. May W. W.: 1975, 'The Composition and Function of Ethical Committees', *Journal of Medical Ethics* **1**, 23–29.
15. *Médecine et expérimentation*: 1982, Cahiers de bioéthique, 4, Centre de bioéthique de l'Institut de Recherches Cliniques de Montréal, Presses de l'Université Laval, Québec.
16. Monod, J.: 1970, *Le hasard et la nécessité: essai sur la philosophie naturelle de la biologie moderne*, Seuil, Paris.

17. National Commission for the Protection of Human Subjects of Biomedical and Behavioral Research: *Report and Recommendations*: *Research on the Fetus*, DHEW (OS) 1976–127; Appendix: DHEW (OS) 1976–128. *The Belmont Report: Ethical Principles and Guidelines for Research Involving Human Subjects*, DHEW (OS) 1978–0012, U.S. Government Printing Office, Bethesda, Maryland.
18. Popper, K. R.: 1961, 'Facts, Standards and Truth: A Further Criticism of Relativism' (First Addendum to Volume II), *The Open Society and Its Enemies*, Fifth Edition Revised (1966), Routledge & Kegan Paul, London, England.
19. Royer, P.: 1982, 'L'éthique et la pédiatrie', *Concours Médical* **104** (25), 4088–4092.
20. Schwartz, D., Flamant, R. and Lellouch, J.: 1970, *L'essai thérapeutique chez l'homme*, Flammarion Médecine-Sciences, 2e édition (1981), Paris; English translation by M. J. R. Healy, *Clinical Trials*, Academic Press, London, England.
21. Veatch, R. M.: 1981, 'Protecting Human Subjects: The Federal Government Steps Back', *Hastings Center Report* **11** (3), 9–12.
22. Villey, R.: 1979, 'La déontologie des essais thérapeutiques', *Bull. Ordre N. Med.* **5**, 205–220.
23. White, S. J.: 1979, 'Statistical Errors in the B.J.P.', *British Journal of Psychiatry* **135**, 336.
24. Willer, J. C., Dehen, H., and Cambier, J.: 1982, 'Analgésie provoquée par le stress, étude électrophysiologique chez l'homme', *Nouv. Presse Med.* **11–18**, 1389–1391.
25. World Medical Association: 1964, Declaration of Helsinki.
26. World Medical Association: 1975, Declaration of Tokyo.
27. Zelen, M.: 1979, 'A New Design for Randomized Clinical Trials', *New England Journal of Medicine* **300**, 1242–1245.

MICHAEL RUSE

AT WHAT LEVEL OF STATISTICAL CERTAINTY OUGHT A RANDOM CLINICAL TRIAL TO BE INTERRUPTED?[1]

One cannot address the question of when to stop a trial without first having dealt with the more basic design questions such as those involved in sample size calculations. But such considerations represent only one aspect of the question. One must also consider the process that must be established in the trial to provide a mechanism that permits one to decide when to stop. . . . The idea that stopping rules can be devised that can be used in deciding when to stop is appealing but it rarely works in practice. Most trials involve observations for various events. It is virtually impossible to prespecify the constellation of data that would be required to terminate the trial.

Statistical analyses can help one in deciding when to stop a trial, but the decision rarely rests on statistical considerations alone. The usual ground rules for levels of significance and values familiar to us in the hypothesis-testing framework do not apply in this decision-making process ([13], p. 632).

There's the rub! And there's the reason for asking someone like me, a philosopher, to enter into a discussion that might seem *prima facie* to be exclusively of concern to medical researchers and statisticians. As the great medical statistician Austin Bradford Hill himself once said:

I have learned that though the statistician may himself never see a patient – although indeed like Tristram Shandy's Uncle Toby he may lead his life in doubt which is the right and which the wrong end of a woman – nevertheless, he cannot sit in an arm-chair, remote and Olympian, comfortably divesting himself of all ethical responsibility ([5], p. 1049).

Suppose that you are faced with some sort of disease, and that you would like to eliminate it, or at least to relieve the suffering that it causes. Suppose also that someone produces a new drug that, for various reasons, they think might indeed go some way towards this end. And suppose further that preliminary studies – for instance, involving the use of animals – suggest that the drug might be effective in the desired way. The next, and probably final step, is to try the drug out on humans in a controlled trial or trials. (For convenience, I shall talk throughout in terms of drugs. But what I have to say is meant also to apply to other medical procedures like surgical innovations.)

At the simplest you will have two groups of sufferers from the disease (*A* and *B*). To the one group you will give the new drug and to the other group you will give nothing, or the best currently available treatment, or some such thing. (There are a number of obvious, much-discussed experimental and ethical questions here which I shall ignore.) After a while, the results start to come in. Specifically one gets reports on the members of groups A and B, which may or may not be similar.[2]

There are literally thousands of examples one could use by way of illustration. At one end, one has limited trials involving just one or two researchers and a few subjects. For instance, Thulbourne and Young (1962) tested the common practice of giving antibiotics to people who have chest surgery. They gave penicillin before and after the operation to 65 patients and gave nothing to 70 patients. They then compared the results (concluding that the drug did nothing to reduce post-operative chest infection nor its severity).

At the other end, one has massive trials, involving many researchers and many subjects. These include the trials of different treatments for pulmonary tuberculosis, which were conducted by the (British) Medical Research Council [8], of cancer chemotherapy by the U.S. Public Health Service [3], and of vaccines against whooping cough [11], TB [10], and polio [15], [10].

Now, there are various ways of analysing the results of trials: ways that use statistical techniques of varying sophistication and complexity. But, essentially what one is trying to see is whether the new drug makes sufferers better, keeps them the same, or has side-effects making them worse (or some combination, like improving the illness but causing new ill-effects). The problem is that, like fire, statistics is a good servant but a bad master. It can tell you all sorts of things about the differences which are emerging in an on-going study, or which have emerged in a completed study. What it cannot tell you is wherein lies the ultimate significance of these differences, or what one should do about them.[3] Nor can it tell you just how certain you need to be that there are real differences, before you decide to act on them.

Specifically, suppose that one were running a trial, and that a difference were starting to emerge. (Alternatively, suppose that one had run one or a series of trials, and differences were making themselves known.) At what point would one want to say: "Enough is enough! Let's start using the drug generally for the disease (or let's forget the drug)"? Clearly, there are all sorts of extraneous (*i.e.*, non-statistical) factors which intrude and which affect one's decision. How desperate is one for a cure? Is the disease very bad? Are there already-existing, reasonable, alternative drugs? Are there any evidences of side-effects, and if so, how bad are they? And so on, and so forth.

These are clearly not simple questions to answer, and yet equally clearly one has got to try to answer them. Moreover, unless one tries to formulate some sorts of guide-lines, one will simply be making intuitive stabs on a case-by-case basis. People are sick. People need drugs. Hence, some decisions have to be made. One cannot continue trials indefinitely.

Furthermore, without some guide-lines, one gets bogged down in a morass of subjectivity. How is one to compare a study done by one set of researchers with a study done by another set? Perhaps one group is claiming great things for its drug A, whereas another group is much more cautious in claims for its drug B. However, then it turns out that the second group is far more stringent in its evaluation of evidence than is the first group. Judged by the same criteria, A and B seem much more comparable, or perhaps even B is better than A.[4]

I cannot hope to answer all of these difficult questions in this paper, nor even to give firm guide-lines. But, I can share some thoughts with you, speaking as a philosopher, about factors which do seem to be pertinent to the problem of what standard of evidence we should demand in trials.

Anticipating my conclusion, what I shall suggest is the following: First, different people have different interests, and these mean that not everyone will ask the same questions of a test or want (or be satisfied with) the same answers. Values enter into thinking about clinical trials. Second, one must beware of a blind, gut demand for overly stringent criteria of evidence. No one wants to prescribe or use worthless drugs. And, with the thalidomide tragedy still very much in our minds, no one wants to prescribe or use drugs which are positively dangerous. Nevertheless, without wanting or pretending to make an absolute case, there are very good reasons why one should not stipulate criteria which are extremely high. Indeed, there may be times – many times – when it is reasonable to accept far-from-definitive evidence. And this could be true of most (if not all) parties interested in the results of a clinical trial.[5]

STATISTICS: SIGNIFICANCE AND OTHER IMPORTANT MATTERS

It would hardly be appropriate for me, virtually a mathematical layman, to recap here what can be found in any elementary book on statistics. And, I am certainly not going to get involved in a technical discussion of rival statistical methods. Let me therefore keep matters to a bare minimum, introducing only enough material to make sense of the more philosophically-oriented discussion which is to follow. (Giere [4] gives an excellent introduction to those aspects of statistics necessary as a background to the philosophical discussion of this essay.)

The situation we are concerned with (at its simplest) involves two groups (randomly chosen and so forth). What we are doing is comparing them, in order to see what sorts of difference (if any) emerge. In some cases, it would hardly seem that anything more than intuition and common sense is needed. Suppose one had two sets of patients (say, each 1000 strong), one set given a drug and the other not. Suppose that all given the drug live, and all not given the drug die. It seems more than likely that taking the drug improves one's health chances! And this holds even if a few given the drug die, and a few not given the drug live. After all, people do die of other causes, and some people have stronger constitutions than others.

Of course, things are usually more complex than this. We often have a varying scale. For instance, rather than simple life and death, we might have months (or years) of life after receiving the drug. Moreover, we will get variation in both groups, and possibly even overlap between the groups. Although the members of group A on average live longer than the members of group B, several members of B may do as well or better than those of A. At this point, untutored intuition collapses! We turn to the statisticians for aid. And in particular, between A and B, the usual advice is that we should apply a *Test of Significance*.

Essentially, what we do here is to try to work out just how typical the two groups (samples) are likely to be of their respective parent populations (*i.e.*, the overall groups from which they are drawn). Then, we estimate just how likely it is that the parent populations are one and the same. In particular, we try to work out the chances that the means (averages) of the two samples being considered are likely to come from the same population or not.

In order to do this, we work from the well-known bell-shaped "normal" curve, which is the curve expected if phenomena are clustered around some mid-point. For instance, in a set of trials of taking the mean of four numbers chosen randomly from 0 to 9, one expects a normal curve around the 4.5 mark (see Figure 1). Even if we are dealing with a population which is not itself normally distributed, we expect a normal curve of means of samples drawn from that population. Moreover, the further we are from the central point of a normal curve, the

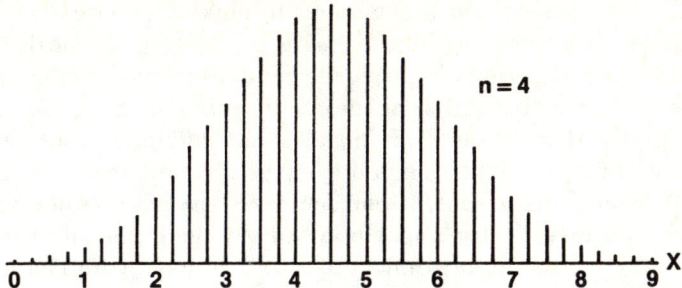

Fig. 1. Distribution of means of samples from a population in which the digits 0 to 9 have uniform probability, for sampe of 4.

less likely it is that a mean will be found there. Hence, if means from two samples fall far apart, it is less and less likely that they really come from the same parent population.

Without going into details, we can quantify the chances that two samples do come from the same population. Most specifically, we can assign probabilities. Is there a 50% chance that two groups come from the same population, or only a 25% chance or a 10% chance? Note that these are only probabilities. By the nature of the case, one can never be absolutely certain.

There are certain cut-off points which are traditionally noted by statisticians and researchers. If there is less than a 5% ($p < 0.05$) that two samples come from the same population, then their difference is said to be "significant". If there is less than a 1% chance ($p < 0.01$) then the difference is "highly significant" (less than 0.1% chance, $p < 0.001$, "very highly significant").

It is important to keep in mind (in fact, the whole point of this paper is to keep in mind!) that these are somewhat arbitrary terms. Because something changes from a 5.1% to a 4.9% chance, our study does not suddenly, magically, become all that better, even though it would now be labelled "significant." Also, the fact that something is significant does not necessarily mean that it is all that interesting. Nor, repeating an earlier caution, does it mean that differences are necessarily *caused* by the things we are studying (especially that the causation is direct, rather than a function of a third phenomenon).

Remember, also, that probabilities are not fixed things, determined once and for all. The nature of the world, the amount of information one has, the extent to which one wants to make a precise assessment, and so forth. Altering any of these makes a difference. In particular, an investigator can attempt to improve the confidence that he (she) has in a claim, particularly by gathering more information. Suppose, for instance, that a drug really does make some difference, but that this difference is rather slight. Limited samples are probably not going to give significant results. But, if one improves the data gathering, then eventually an investigator might hope to get significant answers. The more you know, the more confidently you can make your claims.

One well-known approach to drug testing works on precisely this last-stated principle. In sequential testing, you keep running the experiment until you get significant results. However, this approach is obviously limited in that you need to keep getting back your information fairly rapidly. If it takes a long time to get any results on a test, then it is far less easy to keep up continuous monitoring, until you get some significant answers.[6]

This almost is enough background to go on. Let us turn now to the problem at hand. As must be even more obvious than when we started, asking what level of significance we want from a trial is a bit like asking how tall is tall! However, just as we can say that a tall human is (say) someone over 6′ (and a very tall human, someone over 6′ 6″), so we can analogously be guided by the traditional levels of significance. After all, a less than 1 in 20 chance of error is normally a reasonably good gamble. Moreover, continuing the analogy, just as we can say that probably being much more than 6′ 6″ turns from being desirable phenomenon to a handicap (unless one has aspirations in the direction of basketball!), so we might hope to draw some specific views about desirable levels of significance in trials.

In order to sharpen discussion, let's assume that, all other things being equal, a significant result ($p < 0.05$) is necessary and sufficient. The problem now becomes whether, in clinical trials, one should follow the earlier-expressed intuition and demand a higher level of significance (let us say, at least $p < 0.01$)? Also, are there times when one would accept/be-satisfied-with/feel-immoral-to-insist-on-more-than a result which was less than significant ($p < 0.05$)?

I regard this assumption about significance as something along the lines of what a Popperian would call a "conjecture." It is not something which is true, or perhaps even all that reasonable in some absolute sense. It is a working hypothesis to be thrown out – or at least to be

STATISTICAL CERTAINTY AND TRIAL INTERRUPTION 195

adjusted – if it fails to serve its purpose [16]. Making any such assumption does, however, raise one final question (or set of questions) which must be dealt with in this preliminary discussion, before we can turn to the main topic of this essay.

If we ask for significant results, what are we caring about? What is the most important thing in a test or trial? Is this most important thing the finding out that a drug works positively? Is this most important thing the guarding against a drug working negatively? Is this most important thing the insisting on big differences before we commit ourselves? Or, are we so desperate for a useful drug that we simply can't afford to be too cautious? Can we afford possibly to overlook real differences that a drug brings about, for the sake of safety?

Questions like these raise the area in which statisticians consider the kinds of errors we are liable to make. In particular, it is convenient to distinguish between *Type 1* errors, when we assume that there are differences when there are not (technically, when we incorrectly reject the "null hypothesis"), and *Type 2* errors, when we assume that there are no differences when there really are (technically, when we incorrectly embrace the null hypothesis).

We can draw up a little matrix, distinguishing Type 1 errors from Type 2 errors. Suppose we test a drug (against an older drug), and decide to adopt it. Then we have the four possible outcomes shown in Figure 2.

Decision	Reality (unknown)	
	Old better	New better
Retain old	Correct decision	Error of Type II
Adopt new	Error of Type I	Correct decision

Fig. 2.

The basic paradox you face in statistical testing is the following: the harder you try to avoid committing a Type 1 error, the more chance you run of falling into a Type 2 error! Let me explain through an example. Pretend you have a new drug, given to 100 people, that you have a 50%

recovery rate on the old drug, and that you are looking for better results on the new drug. Your initial position is that you will adopt the new drug, if it does better, but not otherwise (*i.e.*, if the drug leads to anything more than a 50% recovery rate). You can graph the probability of making a wrong decision as shown in Figure 3. Alternatively (and more usually), one can give an "operating-characteristic curve" (OC curve), comparing the new drug against the old drug (Figure 4).

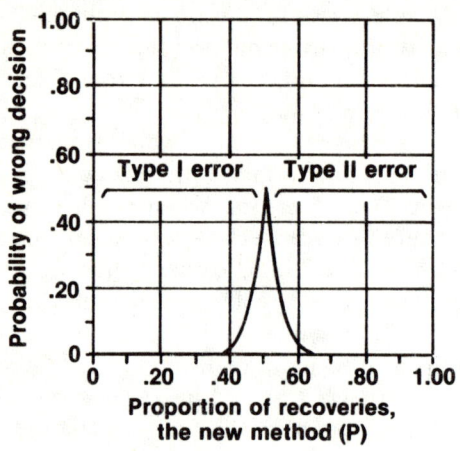

Fig. 3. Probability of wrong decision for various values of P. Decision procedure: Take sample of 100. Adopt new method if $p \geq 0.51$. Retain old method if $p \leq 0.50$. Correct decisions: If $p > 0.5$, adopt new method.
If $p \leq 0.5$, retain old method.

Suppose, now, that you decide to tighten up on the possibility of a Type 1 error. As things stand, even if it's only as good as the old drug, you're laying yourself open to accepting the new drug 50% of the time. Suppose you insist that this isn't good enough. You want to reduce to a 10% chance the possibility of accepting the new drug, if it is no better than the old. One can show, in fact, that this involves pushing the OC curve to the right. Indeed, one can show that the drug should be accepted only if it achieves a 57% improvement rate in the sample (see Figure 5).

Unfortunately, the cost in reduction of Type 1 error is an *increase* in Type 2 error. In fact, if the new drug really gives 60% recovery rate, one

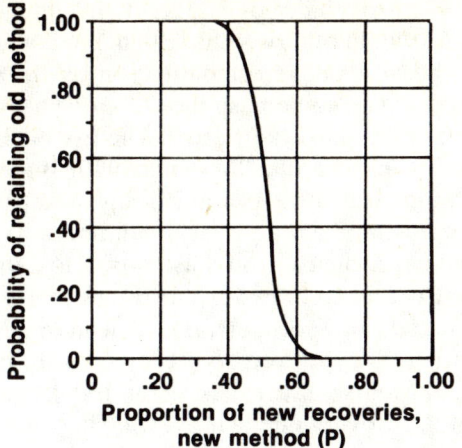

Fig. 4. Operating characteristic curve for test based on sample of 100, old method retained if sample shows 50 or fewer recoveries.

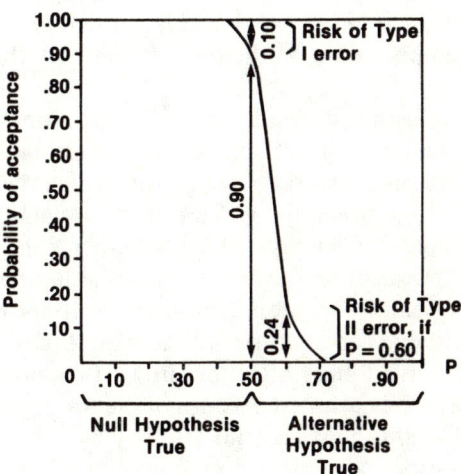

Fig. 5. Probability of accepting the null hypothesis as a function of p ($n = 100$; $p = 0.57$).

now has only a 0.76 probability of finding it. In other words, 10% of one's sufferers, who would otherwise have benefited, will not so benefit. And why? Simply because one rejected the falsity of the null hypothesis.

So what do you do? You could push the OC curve back the other way, thus reducing the chance of a Type 2 error. The trouble is that now what you do is increase the chance of a Type 1 error! In an important sense, therefore, you cannot win. Less pessimistically, what you have got to decide is whether it is more of a threat to commit a Type 1 error or a Type 2 error. And here, clearly, you must appeal to extraneous values.

The main moral I want to draw for our discussion is just how much undue caution (*i.e.*, fear of Type 1 errors) can lead to missed opportunities (*i.e.*, committing of Type 2 errors). For instance, in the just-given drug case, if we put the risk of a Type 1 error at 0.01, then if in fact the new drug had a 60% recovery rate (surely, what in common parlance we would call a "significant" difference above a 50% recovery rate), we would run a 0.69 risk of committing a Type 2 error. Do we really want to be this cautious?

There are escapes. Most obviously, as noted earlier, you can increase your sample size. This means that the chances of making a mistake are reduced. Technically, the *power* of the test has been improved. The trouble is that it costs more in time and labor and so forth to use a bigger sample. Do you have such resources? Is it necessarily going to be worth your while to use such resources on this particular test? After all, everyone has some upper limit on the resource amount that they have to give to testing.

Another help in testing depends on the distinction between "one-tailed" and "two-tailed" tests. Suppose we're testing a drug on 100 patients, against 100 others who do not get the drug. We're looking for significant differences between the two groups. But what differences are we looking for? Are we simply asking whether the drug makes a difference – any difference? Or, are we asking whether the drug actually improves people's health over non-drug taking? Suppose that without the drug people will live (on average) 12 months. Are we asking simply whether people will live longer after the drug – thus taking for granted that no one will have life curtailed because of the drug? Or are we taking the pessimistic possibility into account also?

The former question calls for a "one-tailed test." The latter, for a "two-tailed test." What is important for us here is that the former question, demanding a one-tailed test, asks for less information than the latter question, demanding a two-tailed test. Hence, we can get greater accuracy in our answer for the former question over the latter question, from the *same* initial information. We can either aim for a higher level of significance from the same-sized sample, or use a smaller sample for the same level of significance. The problem, of course, is that we have to be pretty certain that, if there is going to be any difference from our new drug, it will be in the direction of improvement. We're simply ruling out the possibility of its having any dangerous effects.

Enough by way of preliminary. Let us now turn to the main topic. To get to grips with the problems of satisfactory significance tests in trials, it is clear that we have got to think in terms of the people involved (in the loosest possible sense) in a trial. Without pretending to be definitive, there seem to be at least the following parties: medical researchers, subjects in the trial, sufferers from the disease, practising physicians (and/or surgeons), sponsors of the trial, and general public. Some of these may overlap. All of the subjects in a trial may in fact be sufferers from the disease in question(*i.e.*, people who have not been given the disease artificially for the purpose of the experiment). But, for convenience, let us take these groups separately, and ask about their interests, and what effect (if any) these interests should have on our assessment of worthwhile aims in an experiment.

THE RESEARCHERS

I am concerned here with the scientists who discover or identify a drug, and who propose its use in a trial. I am concentrating simply on their own personal interests in the trial. I do not deny that they will have general altruistic feelings of concern towards sufferers from the disease (together with such other feelings, as a concern not to waste their sponsor's money). But these secondary factors are to be ignored here. As promised, I shall be looking directly at the interests of the sufferers shortly. Hence, such items as the empathy that others feel for them can be considered part and parcel of that analysis.

Now, an immediate initial reaction might be that – considered in this restricted sense – the researchers have no interests at all: Their concerns begin and end with the rights and interests of others, namely, the sufferers. This, I think, would probably be the position taken by that sensitive and humane commentator, Hans Jonas. He writes:

In the course of treatment, the physician is obligated to the patient and to no one else. He is not the agent of society, nor of the interests of medical science, nor of the patient's family, nor of his co-sufferers or future sufferers from the same disease. The patient alone counts when he is under the physician's care. By the simple law of bilateral contract (analogous, for example, to the relation of lawyer to client and its "conflict of interest" rule), the physician is bound not to let any other interest interfere with that of the patient being cured. But manifestly more sublime norms than contractual ones are involved. We may speak of a sacred trust; strictly by its terms, the doctor is, as it were, alone with his patient and God ([6], p. 21).

There are a number of comments that will have to be made about this passage, but let me note in passing that – as with all of Jonas's writings – it expresses a sense of the worth and dignity of human beings towards which all of us should aspire. Even if one has something further to add, Jonas's noble sentiments should not be clouded and lost from view.

The obvious point that needs to be made here is that, whatever the truth of Jonas's claim about the physician-patient relationship (of which more later), the claim is not strictly relevant at this point. We are looking at researchers as researchers, and not (as they may well be) at researchers as physicians, caring for and about the sick.

The natural response might be that this gives even less reason (perhaps a counter-reason) for considering the interests of researchers. At least doctors caring for patients have a stake in the well-being of actual flesh and blood individuals, as opposed to researchers manipulating an idealized, possibly unknown class. However, for two reasons, this response is an over-reaction.

First, there is the simple pragmatic fact that, unless you think of the interests of researchers, they are just not going to do the job at all. Certainly, they are not going to do as good a job as otherwise. People get into medical research for many reasons – it is exciting, it is interesting, it holds out prospects of reward (like Nobel Prizes). Whether or not these are dignified ends – and I, for one, would not be so presumptuous as to criticize them – they are very human ends. I doubt very much that any one will alter them in the near future. Hence, just as a matter of simple fact based on human nature, one cannot ignore the kinds of standards of confirmation that medical researchers, as researchers, would like to achieve.

Second, let me make a cry for the value of medical science as science (or as technology, if you like). For all the dangers of science and technology, I believe them to be valuable things in themselves – just as art and music and literature, and even philosophy (!), are. We may be modified monkeys – but, we are yet more than just brutes. We are beings with a power of good and ill – and part of the good is the inquiry into the world, and its control, as manifested through science and technology. Just as we cherish Beethoven's Fifth Symphony for its own sake (and not simply because it keeps musicians in business), so also we should cherish – for its own sake – the knowledge that insulin maintains diabetics. (For more on this kind of point, see Ruse 1981.)

Grant the interests of researchers. How do these interests bear on our title question? Is it in the interests of researchers to aim for high or low levels of significance? (Specifically, in the medical case, should our researcher be satisfied with $p < 0.05$, or should he/she insist on something stronger? And what should he/she be worried about – Type 1 or Type 2 errors?) The answer, of course, is that "it all depends on the particular case." But, let me make the following point. You might think that, as a rule, a researcher would like to be as certain as possible – and this clearly seems to be true. Nevertheless, there are pragmatic factors. If a researcher spends inordinate time and effort on each experiment or trial, he/she will minimize his/her ability to get on and try out many different ideas and hypotheses. He/she may learn much about a little, but he/she will fail entirely to learn a little about much.

In other words, using the language of statisticians, there are costs to increasing the power of a test. If you have only finite resources – time, materials, labor, and so forth – and this ultimately is all any of us have, you must ask how powerful you want to make any particular test, realizing that greater power will entail more stringent economies elsewhere.

I emphasize cost/benefit considerations deliberately, for at the present state of medical research, there are good reasons to suggest that too great a demand for increasingly powerful tests may be highly counterproductive. Let me quote from a recent discussion, trying to pin-point reasons why the discovery of new, useful drugs seems to have slowed rapidly.

One reason for the decline in innovation may be the difficulty of devising suitable laboratory experiments for identifying drugs for the unconquered diseases. When a disease is due to a known infecting organism, compounds can be selected according to knowledge of the biochemistry of the organism, or empirically. Such compounds can be tested against the organism *in vitro* and *in vivo*, and the most effective selected for further study. The search may be long, but the route is clear. It has led to the whole range of chemotherapeutic drugs and antibiotics which have outstanding powers of saving human and animal lives. But this field is largely worked out, except for the tropical diseases. . . . For diseases not known to be due to an infecting organism, some other clue is necessary to start the search for useful new drugs. When a particular substance is shown to be lacking, its replacement is indicated, as with L-dopa for Parkinson's disease and, earlier, in the various endocrine gland failures. New local hormones or mediators continue to be discovered, but it remains to be seen how extensive the practical therapeutic consequences will be of the great body of recent work on prostaglandins, leukotrienes and neuropeptides. . . . The other major strategy for discovering new drugs is to pursue the detail of a physiological or biochemical system and to exploit its practical application. This is the natural course of academic scientists, and has led to such important agents as insulin and demercaprol. In the past thirty years, however, the practical application by academic scientists of their discoveries has declined . . . In a reasonable world, the increasing difficulties of discovery and the seriousness of the incurable diseases may be seen as good grounds for a more tolerant attitude to the hazards of therapy. Desperate ills need desperate remedies, but the wish for completely safe remedies has had and is having the gravest effects on drug discovery. The problem of finding valid laboratory models for the unsuspected toxic effects of drugs is just as difficult as the problem of devising methods for detecting therapeutic potential ([20], p. 388).

Ignoring for a moment the problem of dangers, one moral for the researcher does seem to emerge. All of the "easy" problems in medical research are used up! Work is getting more and more difficult and time consuming. And one simply has to take this fact into account.

What does this all mean in practical terms for the medical researcher, considered *qua* researcher? As pointed out above, you simply have to balance your value options – moving ahead as quickly as possible, given necessarily finite resources. You cannot simply keep increasing the power of a test. But, then you have the question of when you are going to invest in a powerful test and when not, and if not, which is your greater worry: a Type 1 error or a Type 2 error?

Final answers are impossible, unless one goes through instances, case by case. However, there are general guiding principles. First and foremost, matters are going to depend on the extent to which you are working firmly within a well-established theory or research tradition or paradigm. Are you coming at a new test with a body of reliable background information, or not? If you are, then presumably you have certain expectations, already. Probably, therefore, you won't want a very powerful test – at least, not unless something anomalous occurs. Then, you might feel called to recheck in more detail, and with more precision.

Again: prior expectations will influence your decision to guard against Type 1 or Type 2 errors. The point to be kept in mind is one which has been made before. You have a trade-off between the two types of error. In particular, you must beware of taking an overly cautious position, refusing, just as a matter of principle, to reject the null hypothesis, except for truly overwhelming counter-evidence. If your theory gives good reason to think the null hypothesis false, then it is simply bad science to ignore this fact.

In short, demanding super-high significance levels before you will accept real differences, could be counter-productive – and go against the dictates of reason.

Of course, if one is working in an area where less is known, then different strategies may be demanded. If you come up with promising results, then before you commit funds and time to a particular line of research, you may well demand a higher significance level (for instance, $p < 0.01$). Especially, you will be concerned to guard against Type 1 errors. On the other hand, if you're really stuck, looking for any promising avenue, your worries will switch to the dangers of making Type 2 errors.

In short, as one might expect, there simply isn't one straight answer to questions about levels one should, *qua* researcher, demand or accept in testing a new drug. Values, dictated by prior knowledge and future hopes and needs, enter in. A usual significance level ($p < 0.05$) may be a good working hypothesis (in the sense discussed earlier). There are good reasons not to demand too much from a test. But, circumstances will dictate whether one must in fact ask for greater significance levels and what kinds of errors one must guard against. If a result is really strange, and getting the wrong answer would make you look a bit ludicrous, would you honestly want to present it to the world knowing that there is a 1 in 20 chance that you may be wrong?

SUBJECTS GIVEN DRUGS

We must turn next to look at the other people directly involved in a clinical trial, namely, the subjects. I shall assume that we are dealing with people who do have the disease in question, whether they came by it naturally or had it artificially induced.[7] I shall assume also that we are not dealing with a situation which introduces special moral factors (*e.g.*, where people are participating in a trial without their knowledge or consent).

We have two groups of people to consider: those getting the new drug and those not getting it. People may or may not be aware of which group they fall into.[8] There are, I take it, three basic results which might emerge or start to emerge. First, those given the new drug might improve (compared to the control group), with no apparent ill-effects. Second, there might be ill-effects, with no apparent benefit. Third, there might be both improvement and ill-effects. What influence would these results have on the participants' interests in the trial?

Start with the trial where preliminary results indicate that those given the drug do benefit, and where there are no apparent ill-effects. So long as they continue getting the drug, I suppose it really makes little difference to these subjects what level of confidence the researchers aim for. Assuming that the actual participation in the trial causes no great inconvenience, they are relatively indifferent as to whether it goes on a short time or a long time. Their only real concern is that, if the drug does prove beneficial, they should continue to get it once the trial is over. And this is not so much a function of the trial itself, but of the mechanisms for acting on a successful trial.

As far as the actual trial itself is concerned, as always much will depend on the prior knowledge one brings, or does not bring, to it. A major consideration, obviously, is whether one will go for a one-tailed or two-tailed test. In the former case, you simply are not going to get any ill-effects. At least, you are not going to spot them! It is hard to imagine that at no point in testing a drug is the question of ill-effects going to be raised. However, one could imagine a case where the drug's ill-effects are fairly well-known, and new positive information is being sought. Perhaps, for instance, one is trying an old drug in a new way. In various exploratory tests of this nature, one could perhaps reasonably opt for one-tailed tests over two-tailed tests.

What about a trial where there are ill-effects, without any appreciable benefits? Obviously, the answer depends very much on two factors. First, how desperate is one for an effective drug? Is one dealing with a bad illness: one for which there is no known cure? Is one dealing with a wide-spread problem, or one which affects only a few? Second, just how bad are the ill-effects? Suppose one were running trials on the common cold, using hitherto healthy volunteers (*i.e.*, not people who had already suffered greatly). Suppose the effect of the drug were simply to extend the length of the cold from one week to two. I doubt there would be much concern about an extended run of this trial.

Clearly, however, if the drug seems to have very bad effects, one is hardly justified in running the trial for very long at all. I am not sure, indeed, that one would even care or dare to run the trial until one had gotten significant results. I imagine that a decision when to stop such a trial would be predicated on a number of factors. The most important factor would come from the search for causal connexions between the drug and its putative ill-effects.[9] If one were able to make strong causal links fairly quickly, then the rationale for running the trial further would be gone. Conversely, if there were absolutely no known reasons why a drug might have ill-effects (and counter expectations from similar drugs), then one would be correspondingly loathe to call a premature halt to a trial. At least one would push through to a significant result. Presumably, the big worry here would be to avoid committing a Type 2 error. You don't want to conclude that the drug is without benefit, unless there are strong reasons from the test to prove its lack of worth.

Similar considerations and results obtain when one is dealing with the case where a drug apparently shows some good effects and some ill-effects. Much depends on how good the effects are and how ill the effects are. And again, these matters will depend not only on themselves, but on extrinsic factors, like the question of just how desperate we are for an effective drug. We should be on our guard against too-speedy a conclusion to a trial. After all, the whole point of a statistical analysis is to show just what the figures do mean. If we stop a trial before (say) it reaches significance ($p < 0.05$), then, testing against the null hypothesis, we are allowing that there is a better than 1 in 20 chance that our results are not caused by the drug being examined. If the drug is really needed and really effective, then this might not be a risk worth taking, even though the ill-effects appear severe.

Presumably, at this point, one is actively thinking about the power of the test. If an effective drug is really badly needed, then one is going to be prepared to release more resources into testing for it. And clearly, a major worry is about the possibilities of Type 2 errors. Desperate problems require desperate measures. Too much caution is counter-productive.

One question does arise at this point. Suppose some ill effects do start to appear (whether or not there are good effects also). Are we justified *at all* in using human beings in such a trial? Should we not take the most conservative position possible, and halt a trial as soon as there is hint of trouble as an immoral exploitation of human beings?

Although I am sure we can all think of cases where a trial might degenerate into an immoral exploitation of its subjects, ill-effects *per se* do not imply this, nor do they imply that any and every trial should be stopped at the slightest hint of trouble. (I am assuming relatively straightforward situations. Humans are not being exploited to save on animal tests, and so forth.)

Consider matters on the basis of the two great moral theories: utilitarian (teleological) and Kantian (deontological).

From a utilitarian perspective, believing that one ought to maximize happiness and minimize unhappiness, there is a straightforward answer to our problem. Although people in a trial may suffer, this suffering is being done for the general benefit. Undoubtedly, trials as a whole lead to the discovery of effective drugs and to the avoidance/elimination of ineffective/dangerous drugs. Hence, a certain amount of suffering of individuals can be accommodated, within one's moral norms. Such suffering may indeed, as in the case of trials, be expected. Its probable occurrence is no reason to ban trials.

The Kantian – demanding that people be treated as ends and not as means – has rather more trouble with the dangers of trials, although the problems are not insuperable. Clearly, trials cannot be run simply to satisfy the Nobel-Prize aspirations of the researchers (not that utilitarians would think much of this situation either!). But, no Kantian would deny the right of some people to put themselves at risk – both for the sake of benefits which they themselves might gain, and even more for the sake of other sufferers. In the case of voluntary participation in a trial, the subject is not just a means to the researcher's ends; together with the researcher, he/she is a person in his/her own right, working towards a common goal. (Non-voluntary participation is, of course, another matter, and causes major problems for the Kantian – even if there were no other way in which the trial could be run. I sense that a Kantian might have trouble with the use of prisoners in tests, no matter how "voluntary" their participation. Apart from the use of prisoners for other ends, no matter what they might say, I'm not sure a Kantian would approve of "buying off" punishment.)

The same conclusion, that purely voluntary participation in a trial is acceptable, seems to follow from the ideas of contemporary Kantians, for example, those of John Rawls in his celebrated *A Theory of Justice*. Pushing the notion of "justice as fairness", Rawls would have us make decisions as though from behind a veil of ignorance.

> The idea of the original position is to set up a fair procedure so that any ideas agreed to will be just. The aim is to use the notion of pure procedural justice as a basis of theory. Somehow we must nullify the effects of specific contingencies which put men at odds and tempt them to exploit social and natural circumstances to their own advantage. In order to

do this I assume that the parties are situated behind a veil of ignorance. They do not know how the various alternatives will affect their own particular case and they are obliged to evaluate principles solely on the basis of general considerations ([17], pp. 136–137).

In fact, this is not simply metaphorically but literally the situation we have in a clinical trial. You might benefit; you might suffer; you might stay the same. But there is no immorality here, if you have made a free decision to participate, in the light of information that the researcher has prior to trial. (That is, ignoring such cases as those in which the researcher deliberately conceals risks.)

SUBJECTS NOT GIVEN DRUG (*i.e.*, CONTROLS)

Let us turn now to the case of the subject in a trial who is *not* receiving the new drug, the person who (knowingly or unknowingly) remains on more conventional treatment. Essentially, we have the reverse to that of the subject on the new drug. If the drug proves ineffective or dangerous, then the control subject has no direct interests. The trial can go on for either a short or a long period of time. But if the drug proves effective, then it is in the control's interests that the trial not be too long, so that he/she can also benefit from the new drug.

Nevertheless, as in the case of the person getting the drug, morality does not require that, as soon as the drug has good effects, the trial be ended so that the controls can gain possible benefits. Both utilitarian and Kantian agree that the researcher has the right (if not obligation) to go ahead and finish the trial, without being forced to draw premature conclusions. One has the right – if not the obligation – to push through to significant levels (at least $p < 0.05$).

I suspect that not everyone would agree with me here. Consider the worries of Paul Meier, Isham Professor of Statistics at the University of Chicago:

I was close to those who were knowledgeable and concerned when one of the earliest clinical trials of the Veterans Administration was carried out – testing the efficacy of streptomycin in the treatment of tuberculosis. The study called for the recruitment of, I believe, 200 tuberculous veterans who were moved to Washington, D.C., for a year's therapy and study. Half were given the "best current therapy," and half got that plus streptomycin. At the end of the year, streptomycin was a clear winner, and it was given to all the survivors of the trial.

No one seemed very concerned about possible ethical problems at the time. Streptomycin was in very short supply when the study began, and, but for the study, none of the 200 men could have expected to get it then. But 6 months later, streptomycin was more readily available and I found myself troubled by the ethics of continuing to withhold it from the control subjects. To be sure, we learned a good deal more about the effectiveness of streptomycin in various stages of tuberculosis during that period, but were the patients being dealt with fairly? (quoted in [12], pp. 633 – 640).

Also, let me refer again to Bradford Hill:

When, in 1946, the medical research Council's Streptomycin in Tuberculosis Trials Committee set out to investigate the effect of that drug in pulmonary tuberculosis it was faced with no serious ethical problem. The antibiotic had been discovered two years previously, its striking powers *in vitro* and in experimental tuberculous infection in guinea-pigs had been reported; the published clinical results were distinctly encouraging though not conclusive. Yet overriding all this evidence in favour of the drug was the fact that at that time exceedingly little of it was available in Great Britain, nor were dollars available for any wide-scale purchase of it from the U.S.A. Except for that situation it would certainly on ethical grounds have been impossible to withhold the drug from desperately ill patients. *With* that situation, however, it would, the Committee believed, have been unethical *not* to have seized the opportunity to design a strictly controlled trial which could speedily and effectively reveal the value of the treatment. There was no dearth of patients with the type of disease defined (acute progressive bilateral pulmonary tuberculosis of presumably recent origin, bacteriologically proved, unsuitable for collapse therapy, age-group 15-30).

There was no possibility of obtaining sufficient streptomycin for them all. There was no other suitable form of treatment for them but bed rest.

Thus, knowing that all the streptomycin available in this country was being effectively used – much of it for two rapidly fatal forms of the disease, the miliary and meningeal – the Committee (1948) could proceed not only without qualms of conscience but with a sense of duty to do so ([5], p. 1044).

As Hill points out, because of the limited availability of streptomycin, the proper ethical route was straightforward. But, what if streptomycin had been readily available as started to be the case in the test Meier describes? Hill argues that although "the published clinical results were distinctly encouraging though not conclusive," had the drug been available "it would certainly on ethical grounds have been impossible to withhold the drug from desperately ill patients." In other words, great statistician though he may be, Hill rather suggests that there are times when you need not even achieve significance!

Perhaps we just have different ethical intuitions here: but let me say two things in my own defense. First, in a case like this, the usual form of clinical trial may be inappropriate. You already know part of the answer! Without the drug, you will be badly. Hence, a trial can be run, at least particularly, *retrospectively*. The control group could be drawn from people who suffered (and perhaps died) before the trial was run. In the trial itself, the drug is given to as many sufferers as possible. Thus one avoids the dilemma of deliberately withholding a potentially useful drug.

Second, however valuable a drug may seem, if you really do have genuine doubts about its effectiveness, you must recognize the dangers you run of raising false hopes by administering a worthless remedy (in other words, of committing a Type 1 error). And this could lead to worse consequences: for instance, people might drop already-existing limited therapies, when beguiled by the new "cure." You are therefore not simply faced with the choice of giving/not giving a drug to sick people. You could also be making sick people worse. Hence, the person who insists on pursuing a trial does have some not-obviously-irrelevant points to back him/her up. (In the final sentence of the above-quoted passage, Meier hints at the value of continuing such a test.)

Again, let me emphasize that no tests today are occurring in a vacuum. We know a great deal about the effects of drugs. And one thing which we certainly know is that new "wonder" drugs have a nasty habit of turning out not to be so wonderful after all. Certainly, if you have some drug which is reasonably effective, there are reasons to beware of making Type 1 errors. There are reasons not to rush into rejecting the old, almost just because the new is new. Here, as always, you must juggle the various values. How effective is the old treatment? How much is the new treatment needed? And so forth.

Let me note additionally that there is more to testing a drug than running a clinical trial. You have *in vitro* tests, tests on animals, general background knowledge, and so forth. Factors like these could lead to the rational decision that it is simply not necessary – especially in an emergency – to run a trial, until one gets a significant result. You already have enough information to make a rational decision. In fact, Hill's account rather suggests that this may have been the case with respect to streptomycin in 1948. In this case, he and I have no quarrel. (I suspect, however, that we do still disagree. If one did have enough information, why then was the trial so valuable?)

Be all of this as it may, normally one will want to achieve at least significance in a trial, specifically aiming at minimizing the chances of Type 1 errors. But, concluding both this and the last section, from the viewpoint of subjects, we have seen that there are good reasons not to insist on pushing trials to super-high levels of significance, or of unreasonably demanding too overly powerful tests. If the effects are bad, those receiving the drug should be excused as soon as possible. If the effects are good, those receiving the drug should be included as soon as possible.

SUFFERERS

I turn now to the general case of those suffering from the disease in question. What are their special interests in a clinical trial of a drug which might have healing effects? How do these interests affect the significance levels that researchers should aim for?

Obviously, much that has been said in the last two sections, particularly about control subjects in a trial, applies directly here. Most specifically, if a drug looks as though it is doing some good, it is in the interests of the sufferers from the disease that these beneficial effects be confirmed, and the drug made generally available, as soon as possible. Of course, as before, how great an interest the sufferers have is a function of the severity of the disease, the availability of alternative remedies, and so forth.

However, one might feel that a note of extreme caution should be struck. Although we may want an effective drug, as soon as possible, what we really want is an effective safe drug, as soon as possible. Thalidomide looked as though it were a safe tranquillizer, and look where that got us. Hence, from the perspective of the general sufferer, it would seem that one really ought to run trials for a very great length of time, until one is very, very sure that one is dealing with a safe drug. Very high levels of significance seem to be demanded (*i.e.*, at least $p < 0.01$, if not even more stringent), in order to avoid missing bad effects. And one should check that drugs are safe, not just over the short time, but over the long term. Morality and common sense both dictate this. After all, we all properly have a feeling that sins of omission are not quite as heinous as sins of commission. We may be keeping people from an effective drug. More importantly, we are also keeping them from a potentially dangerous one.

I am not convinced by this line of argument. Intuitions aside, omissions seem just as bad as commissions. Type 1 errors can be as bad as Type 2 errors. A very strong case has got to be made for refusing a desperately ill person a drug which seems to have a very good chance of help, particularly if the only contrary case is for an as-yet-unknown hypothetical ill-effect.

Moreover, as far as ill-effects are concerned, it is important to keep matters in perspective. Probably no drug is absolutely safe or has absolutely no ill-effects on anyone. This applies even to such widely used non-prescription drugs as aspirin. Furthermore, no trial is going to rule out all possible side-effects. Apart from anything else, how is one to deny that, at some point down the road, the drug in use might turn nasty. Perhaps one tests a drug for five years, and then, five years later, when it is in general use, its first users start to develop cancer. Hence, even at best, one is working with probabilities.

Bearing out what I am saying, consider the following two (typical) quotes made by the U.S. Food and Drug Administration about oral contraception.

It is to be emphasized that all known human carcinogens require a latent period of approximately one decade. Hence any valid conclusion must await accurate data on a much larger group of women studied for at least 10 years (quoted in [7], p. 755).

And:

Much indirect evidence suggests steroid hormones, particularly estrogen may be carcinogenic in man. These data are derived from experiments on laboratory animals in which long term administration of estrogen resulted in cancer in five species. Although all physical and chemical agents that are carcinogenic in man produce malignant tumours in experimental animals also, evidence of carcinogenicity of estrogen in other species cannot be transposed directly to man. Suspicion lingers, however, that the result in laboratory animals may be pertinent to man. Many difficulties arise in the epidemiological elucidation of this suspected relation. The principal obstacle is the long latent period between the administration of a known carcinogen and the development of cancer in man. Thus far, no properly devised perspective or retrospective studies have provided an adequate solution to this problem (quoted in [7], p. 763).

Is one to wait at least ten years after every drug trial because of possible cancerous effects?

Of course, one should not now drop straight into the fallacy of black-and-white thinking. That there will always be some risks does not mean that one should not strive to eliminate risks. But it does mean that, always, one must make a judgement call about when to end a trial. Nature does not do it for you!

Furthermore, as always in life, one must not become paralyzed because of potential dangers. I might have a fatal car accident when driving from Guelph to Toronto (60 miles) – people have had such accidents. But I still go on driving to Toronto, because I judge the benefits to outweigh the risks. Similarly in drug trials. If one wants effective drugs badly enough, then some risks must be taken. To requote a passage given earlier:

In a reasonable world, the increasing difficulties of discovery and the seriousness of the incurable diseases may be seen as good grounds for a more tolerant attitude to the hazards of therapy. Desperate ills need desperate remedies, but the wish for completely safe remedies has had and is having the gravest effects on drug discovery. The problem of finding valid laboratory models for the unsuspected toxic effects of drugs is just as difficult as the problem of devising methods for detecting therapeutic potential.

Furthermore:

The major tragedies which have created public alarm and fear and which have led to the condemnation of drugs have arisen in special circumstances (pregnancy and thalidomide; consumption of amine-rich foods and monoamine oxidase inhibitors; Japanese life and clioquinol) or have been due to unexplained sensitivity in a minority of patients (triparanol, practolol, clozapine). Practolol, a B-blocking agent introduced for the control of cardiac arrhythmias and subsequently found valuable also in hypertension, is particularly notable for the thoroughness with which its toxicology was studied in animals, to the satisfaction of registration authorities. Nevertheless, adverse reactions, some serious, developed in a small proportion of patients who received the drug; the ill-effects included skin rashes, impairment of secretions and, most gravely, corneal damage sometimes leading to impairment or loss of vision (the "oculomucocutaneous syndrome").

None of the recognized laboratory procedures gave warning of the serious clinical syndrome nor, even with hindsight, has a procedure been discovered which would have predictive value. Other "idiosyncrasies" are associated with particular genetic factors, as with reactions to suxamethonium, primaguine, penicillamine, sodium aurothiomalate and hydralazine. When much more is known about individual biochemical variation and its genetic control, it may be possible to predict that particular drugs will be hazardous to particular individuals. But the animal models of orthodox toxicity testing give no basis for such subtle predictions ([20], p. 388).

There are risks involved. But consider also the other side of the story:

Delay in the introduction of suitable anti-arrhythmic agents has been estimated to cost 10,000 deaths a year in the United States. Other substantial delays in introducing new drugs into the United States as compared with the United Kingdom have been carefully documented: 7 years for the first B-blocking agent for hypertension, 5 and 3 1/2 years respectively for cromoglycate and inhaled beclomethasone for asthma, 5 years for contrimoxazole for urinary infections. It is difficult to estimate in every instance how many episodes of illness would have been prevented or treated more effectively if the drugs had been available. But presumably a sufficient quantity of evidence has accumulated in each case to justify the widespread use of the drugs. Why else did the Food and Drug Administration (FDA), in the end approve their use? [20], p. 390).

Surely, from the perspective of sufferers, one should be wary of asking for abnormally high safety readings from clinical trials? [10]

Let me repeat a point which has been made before in this essay. We don't live in a perfect world, with virtually unlimited resources of time and help and money. All of the time, value decisions enter in. This means that you have to go beyond mathematics. To say that you want a significant result (or a highly significant result) in a test, just does not make any sense, if just said in isolation. To make sense of the statement, you know just what you are testing for, and how great a proportion of resource material (thus affecting the power of the test) you are prepared to commit.

Is your main concern that of finding useful new drugs? Set the test to avoid Type 2 errors. Is your main concern that of not getting involved in less-than-useful drugs? Set the test to avoid Type 1 errors. Is your main concern to avoid ill-effects? Set the test accordingly – don't use only one-tailed tests. Must you be really sure? Are you dealing with a disease which affects many? Put resources in to increase the power. And so forth. According to your needs, you will have your values, and set your priorities.

I emphasize that there is no unique correct answer to the ends of statistics in testing. And specifically here, considering the interest of sufferers, one should not automatically assume that safety is the one deciding value-factor. Here, as elsewhere in life, taking risks is not necessarily the irrational course of action.

PHYSICIANS

Where do physicians stand with respect to clinical trials? This question, of course, presupposes some answer to the prior question: What is the physician's status, or more particularly, to whom is the physician obligated? As we have seen, Jonas argues for an extreme position. The physician is obligated to and only to his patient. "The patient alone counts when he is under the physician's care" ([6], p. 21). I suppose, if one took this position to the extreme, one could argue that in certain circumstances it would be acceptable (from the physician's perspective) for some clinical trials to go on virtually indefinitely. In such a way, maximum information would be available at precisely the moment the physician needed it. The suffering and inconvenience of anyone other than that particular patient is irrelevant.

This is ridiculous! Even ideally, the physician/patient relationship does not exist in a vacuum. Although a physician's direct concern may be with some particular patient, he/she has duties to his/her other patients (past, present, and future), and in a more general sense to other humans, especially the sick. (In fact, Jonas himself rather concedes this, for he does acknowledge and accept the right to impose quarantine, which is something occurring for the sake of others.)

This being so, I suspect that a physician's interests in a clinical trial are almost purely a function of the interests of the general sufferers from the disease in question. The physician wants safe, effective drugs, no more and no less than does his/her patient, or anyone else similarly afflicted. Thus the conclusions drawn in the last section apply directly. Most particularly, we have the applicability of conclusions about the need to balance factors of safety against factors of effectiveness and need.

No physician wants to give a patient a drug which does not work. But, it is not necessarily a violation of the physician/patient relationship to take a chance with a drug. At times, it may be worth using a drug, even though there are questions about its safety – rather than using nothing at all. Hence, here too we have factors warning against demands for too high a level of significance. Especially, we have warnings against too stringent avoidance of Type 1 errors, comparing a drug positively against the null hypothesis, and against putting too much effort into testing for dangers as against testing for virtues.

SPONSORS

Generally, I take it that sponsors will be one of two kinds (or a combination): publicly funded and privately financed. If the sponsors are publicly funded, then ultimately their responsibility – through various levels of politican and civil servant – is to the general public. If the sponsors are privately financed, then ultimately their responsibility is to their shareholders. Drug firms exist to make money.[11] Let us leave the case of the general public until the next, final section, and concentrate here on trials conducted by private firms.

The aim of a firm is to manufacture drugs, which it can then sell at a profit. Leaving aside questions of monopoly, a drug firm is facing competition from other firms (at home and abroad) which similarly want to sell drugs. Obviously, therefore, it is not in any firm's interests to have indefinitely long trials. Otherwise, it will never get its drugs to market. Nor do I think it fair for society to insist on abnormally long trials, in the name of safety. If we let private business enter into drug search and manufacture, then we must not impose impossible standards.

At some point, the doctrine of *caveat emptor* has to be allowed to come into play. After all, analogously, we do not insist on abnormally high standards of safety from the automobile industry. Why then pick out the pharmaceutical industry for harsh treatment?

On the other hand, this does not mean that private firms will want virtually no trials at all, or that society has no right to insist on any such trials. John Stuart Mill ([14], p. 151) thought that, in the name of liberty, anyone should be allowed to sell virtually anything he wants. Today we accept (rightly) that society can impose quite stringent regulations on private industry – and this certainly extends to the right to insist on some safety precautions on drugs.

Furthermore, it is obviously enlightened self-interest for firms to run trials, checking both the effectiveness and safety of their products. No firm wants to produce another thalidomide. It is disastrous both to reputation and to profits (at least, it ought to be!). Any car firm which has a bad safety record is going to suffer in the market place. The same holds of a drug company.

The conclusion, therefore, points to trials, but not to indefinite lengthy ones, demanding very high levels of significance. You do not have a license to insist that drug firms simply keep raising the power of their tests. If you are prepared to let enterprise enter the drug field, you cannot then ignore or suspend all of the usual rules and practices which apply elsewhere in a capitalist society.

THE GENERAL PUBLIC

The average citizen seems to have two interests in drug trials. First, he(she) is concerned inasmuch as he(she) is sponsoring such trials. Second, he(she) is concerned as a potential beneficiary of such trials (personally, or through a family member, or the like). As a sponsor, one may not be after a direct profit, but points made in the last section are certainly not entirely irrelevant. In particular, one wants to get reasonable value for one's money. The ability to pay taxes is not unlimited, despite contrary views in some quarters.

This all being so, one's interests are certainly not in trials which simply go on and on, forever raising the power of the tests. One will want compromises, balancing the availability of a range of drugs, safety, costs, and like factors. Take an analogy. In Ontario, electricity is publicly financed. Balances are struck between the need (or wish) for lots of reasonably priced power, the costs of new generating stations, and drawbacks such as pollutants (leading to acid rain) and the escape of radioactive material. Similarly in the production of drugs. If we are prepared for substantially higher costs, then bigger and perhaps longer trials can be run – leading to greater levels of significance. By increasing the power of tests, we can cut down on the chances of error. But, it is by no means obvious that this is a process which can or should go on indefinitely. After all, there are desirable things in life, other than yet more drugs.

And, this point surely holds for the general public considered as potential consumer. When you fall sick with something, then what you want is a safe, effective drug. But, particularly given what has been said earlier about the need to compromise between undue safety restrictions and availability of a wide range of remedies, no one can rationally ask that trials go on virtually indefinitely. If I fall ill with something, then I need help. It is not much joy to be told that a new drug exists, but that it is still under test. (I am not now suggesting that no tests should be run!)

CONCLUSION

At the beginning, I suggested that one's intuitive feeling – certainly my intuitive feeling! – is that clinical trials should be run until researchers achieve very high levels of significances using very powerful tests. After all, one is dealing with human beings – sick human beings – and it seems, therefore, wrong to take any chances. What I hope I have shown in this paper is that matters are nothing like straightforward. Clinical trials are essential in the testing of drugs – for effectiveness and for safety. But there are lots of different values and interests which enter into testing, and these are no less important than the pure mathematics of statistics.

I feel that this essay is more of an exploration of topics not much (regretfully not much) discussed by philosophers, than a set of definitive pronouncements. Hence, I am hesitant (and indeed unable) to draw any final conclusions. But there are two major points which emerge from the discussion. They are both worth emphasizing.

First, we have a lot of different people interested in the results of clinical trials. And it is naive to pretend that they will all be interested in exactly the same way. Put matters at the bluntest. A researcher could be attracted to a particular area, because there are great prizes for success.

And, there may well not be equal penalties for dangerous failures. To take a topical matter, think about herpes. No argument is needed to suggest that the person who comes up with an effective cure is going to reap maximum professional rewards. Moreover, producing a drug with some unfortunate side-effects need not be professionally devastating – after all, the scientific community itself will not necessarily be adversely affected.

However, another person interested in a herpes drug trial may not have the same values. A physician, for instance, may feel that although today herpes is a high-profile disease, it is really something the average person can live with without too much inconvenience. Its unpleasantness does not warrant serious life-threatening risks. Hence, the physician has different stakes from the researcher. The researcher is eager to eliminate Type 2 errors, leading away from promising drugs. The physician is concerned, among other things, to avoid Type 1 errors, involving assumptions that a drug has benefits when it doesn't really.

And others will have yet other interests. Consider, for instance, the as-yet mysterious disease, AIDS ([1], p. 749). Researchers and physicians will come together on the value of reasonably powerful HIV diagnostic tests. So also will those in the general population who are at high risk. But, those concerned with overall health-care costs may (foolishly?) decide that, although dreadful, AIDS is far too restricted a disease to merit full-scale, emergency support. Better by far to invest in lower-profile, more prevalent ailments, they might argue.

I am not suggesting that compromise is not possible. It has to be! I am suggesting that compromise may be necessary. Different people have different interests, and these will affect the questions asked and the answers found satisfactory.

The second point arising from the discussion qualifies but in no way eliminates the first point. One should not assume automatically that it is in everyone's interests to maximize the probability of error, or specifically, to maximize the cautious strategy. Nor should one assume that caution will separate one group, researchers from other groups, particularly sufferers. We have seen arguments to suggest that sufferers themselves may not always endorse the most-cautious approach to drug testing. As always when values are concerned, balances must be struck. High significance levels guarding against dangerous drugs may look desirable. But, there are costs. When helpful drugs are desperately needed, a willingness to take risks may well be the best policy.

In short, you simply can't give one answer to questions about levels of significance in clinical trials. But that doesn't mean that all is totally subjective. As this essay has tried to show, here as elsewhere in life where values are concerned, reason can produce some useful guidelines.

University of Guelph
Guelph, Ontario, Canada

NOTES

[1] I should like to thank Tristram Engelhardt and Ronald Giere for critical comments on an earlier version of this paper.

[2] This is all very simplified! At least, one will have stages, probably requiring a sequence of trials. And the trial itself is but one part of drug testing.

"A carefully designed controlled trial . . . is of course only one link (although a highly important link) in a chain of investigations about specific medical treatments. A new drug, for example, will undergo studies in animal and human pharmacology to elucidate its mode of action and to detect any signs of acute toxicity. It will probably then be used tentatively on a small number of patients before a full-scale clinical trial is contemplated. In these early studies on patients it is customary to refer to a *Phase I study*, purpose of which is to search for non toxic dosage schedules; some writers include the first human pharmacological studies (usually on healthy volunteers) in this category. A *Phase II trial* is the first attempt to access therapeutic effect. It can be regarded as a screening procedure; unpromising drugs are dropped at this stage whilst more promising ones continue to Phase III – the full-scale trial. Even after a Phase III trial a need may be felt for more experience about a particular form of treatment in the hands of non-specialists, for it can be regarded as a proven value. Such follow-up studies, often conducted by a wide range of medical practitioners, are sometimes called *Familiarization Trials*" ([2], pp. 6–7).

[3] For instance, male homosexuals tend to report closer bonds with their mothers than do male heterosexuals. Does this prove the Freudian contention that dominant mothers tend to cause male homosexuality or that (knowing about Freud) male homosexuals give the "proper" answer, or that being homosexual makes you more fond of mum? And what, if anything, should you do with the information?

[4] Consider the putative link between oral contraception and an increased risk of thrombosis.

"Studies conducted in Great Britain and reported in April, 1968, estimate there is a seven to tenfold increase in mortality and morbidity due to thromboembolic diseases in women taking oral contraceptives . . ." ([7], p. 763).

But, do these studies tell us anything about the American scene? The following revealing interchange occurred before a U.S. Senate Committee.

"SENATOR MCINTYRE: Dr. Wynn, the statement has been made here several times that the British findings regarding increased risk of thromboembolic disease with use of the pill cannot be applied directly to women in other countries including the United States. In fact, a statement to this effect was allowed in the official labeling of the oral contraceptives by FDA in 1968.

Moreover, the Sartwell study did come up with a different finding regarding the magnitude of the increased risk in the United States as opposed to Great Britain.

Do you know, Doctor, of any reasons why the British and the American female population should be any different with respect to the increased risk of thromboembolic disease resulting from use of oral contraceptives?

DR. WYNN: Taking the women by and large, I think the answer to that question is no. I would not expect there would be substantial differences between British and American women so far as risk of thromboembolism is concerned.

What the data reveal is the very great difficulty of carrying out epidemiological studies. You see, you take the Sartwell study. It was not identical to the British study.

They identified over 2,500 cases – I am speaking from memory – 2,500 cases of thromboembolism, but they excluded all but a small fraction, 176, for one reason or another. Now some of the reasons for exclusion were in my view unreasonable.

They excluded women with varicose veins and family histories of diabetes, and so on. Now in conversation with Dr. Sartwell, he has agreed with me that some of the reasons for exclusion were not really those which he would advocate at the present time. Be that as it may, it merely gives some indication of the great difficulties in this type of study" ([7], pp. 769–770).

[5] I am not, of course, suggesting that once a trial is over, one should thenceforth have a closed mind. General use of a drug gives new evidence, which then gets fed back into one's position. My concern is with the initial decision about whether or not to go ahead and use the drug.

[6] An example of a sequential trial was one performed by Snell and Armitage [18] designed to test cough suppressants.

"There were three treatments, heroin (diamorphine), 'lipect' (pholcodine) and a placebo, and these were administered in a random order to patients with chronic cough. Each treatment was prepared in the form of a linctus. Each patient received a particular linctus on two successive evenings, and at the end of 6 days (having received all three treatments) he was asked to rank them in order of preference. Some patients were unable to state a preference between some or all of the treatments, but a separate chart was used for each pair of treatments, and all available preferences were plotted sequentially. The total time span of the trial was fixed by other commitments. The rate at which preferences were becoming available was estimated from the first part of the trial, and this enabled a reasonable guess to be made of the total number of preferences which would be available for each comparison, in the maximum allowable time" ([2], pp. 60–61).

The results can be charted as given in the following figure:

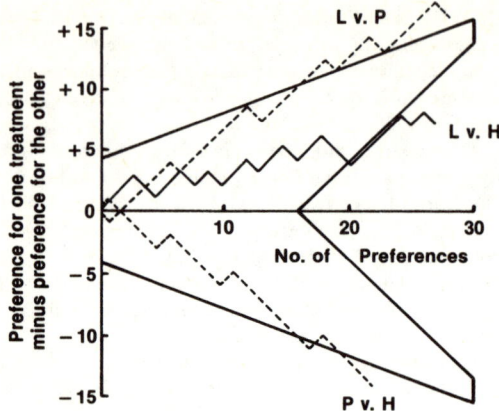

Chart for sequential analysis of results: H, heroin; L, lipect; P, placebo. In each comparison the path is continued one unit "northeast" for a preference for the first-named treatment, and one unit "south-east" for a preference for the second-named treatment. The upper and lower bounds give the points at which one could say that there were significant differences ($p < 0.05$) between drugs. The center bounds give the points at which one could say that there would be no such differences. As you can see, after 17 results one could say that "lipect" was better than the placebo, and the same holds for heroin compared to the placebo. After 20 results, one could say that there was no significant difference between "lipect" and "heroin."

I must repeat that one cannot always use sequential methods.
"Sequential methods are therefore particularly suitable for trials to assess the treatment of acute conditions, or those in which rapid relief of a chronic condition is sought. Trials of long-term treatment of chronic disease, where the measurement of response can be made only after a long follow-up period, are much less amenable to sequential methods" ([2], p. 26).

The question posed in the title of this paper is somewhat misleading, in that it rather suggests that all trials are (or could be) sequential. Really, the question to be asked is "What level of statistical significance is satisfactory or adequate?" This clearly applies to all trials.

[7] This is not too great a restriction and saves a lot of qualification. Things that I have to say about sick subjects can be modified to apply, where appropriate, to healthy subjects.
[8] Sometimes it is important that people not know where they themselves stand in a trial. Otherwise, psychological perceptions might distort the results. At other times, I presume, it would be difficult to avoid release of some information. Suppose, for instance, one were trying out a new kind of surgery. I imagine that those who had it would usually know that they had had it. Whether those who did not have it would know that they did not have it, is perhaps another matter. It is surely plausible to suppose that there would be times when they would know that they had not had the surgery in question.

⁹ Remember, a trial does not lead directly to causes. A trial gives correlations. From these, if we can, we infer causes.

¹⁰ As I am sure you will have realized, the problem we are faced with here is not simply one of significance levels. It is also a function of what you are testing for. If you are satisfied with a subject being cancer-free 10 minutes after a drug is given, then you will get results much more quickly than if you insist on studying subjects for 10 years after drug use. But significance levels do enter into the problem. The longer you take, the more extraneous factors affect people's health, and the more tests you must make – and subjects you must study – to pin-point correlations with the same level of accuracy.

¹¹ I am trying to state a fact, and not to make a moral claim here. Whether or not people should be allowed to make a profit out of sickness is another matter. A simple fact is that we in the West accept that free enterprise can properly move into the realm of health care, and an equally simple fact is that the ultimate goal of a private business – any business – is to make money.

BIBLIOGRAPHY

1. Anonymous: 1983, 'No Need for Panic about AIDS', *Nature* **302**(28), 749.
2. Armitage, P.: 1975, *Sequential Medical Trials*, 2nd ed., Wiley, New York.
3. Endicott, K. M.: 1957, 'The Chemotherapy Program', *J. National Cancer Institute* **19**, 275–93.
4. Giere, R.: 1979, *Understanding Scientific Reasoning*, Holt, Rinehart and Winston, New York.
5. Hill, A. B.: 1963, 'Medical Ethics and Controlled Trials', *British Medical Journal* **1**, 1043–1049.
6. Jonas, H.: 1970, 'Philosophical Reflections on Experimenting with Human Subjects', in P. A. Freund (ed.), *Experimentation with Human Subjects*, George Braziller, New York.
7. Katz, J.: 1972, *Experimentation with Human Beings*, Russell Sage Foundation, New York.
8. Medical Research Council: 1948, 'Streptomycin Treatment of Pulmonary Tuberculosis', *British Medical Journal* **2**, 769–782.
9. Medical Research Council: 1955, 'Various Combinations of Isoniazid with Streptomycin or with P.A.S. in the Treatment of Pulmonary Tuberculosis', *British Medical Journal* **1**, 435–445.
10. Medical Research Council: 1956, 'B.C.G. and Vole Bacillus Vaccines in the Prevention of Tuberculosis in Adolescents', *British Medical Journal* **1**, 413–427.
11. Medical Research Council: 1959, 'Vaccination Against Whooping Cough', *British Medical Journal* **1**, 994–1000.
12. Meier, P.: 1979, 'Terminating a Trial – The Ethical Problem', *Clinical Pharmacology and Therapeutics* **25**, 633–640.
13. Meinert, C. L.: 1979, 'When and How to Stop a Clinical Trial', *Clinical Pharmacology and Therapeutics* **25**, 632.
14. Mill, J. S.: 1859, *On Liberty* (Reprinted 1910), Dent, London.
15. Poliomyelitis Vaccine Evaluation Center: 1955, *Evaluation of 1954 Field Trial of*

Poliomyelitis Vaccine, The National Foundation for Infantile Paralysis, Ann Arbor, Michigan.
16. Popper, K: 1962, *Conjectures and Refutations*, Basic Books, New York.
17. Rawls, J.: 1971, *A Theory of Justice*, Harvard University Press, Cambridge, Massachusetts.
18. Snell, E. S., and Armitage, P.: 1957, 'Clinical Comparison of Diamorphine and Pholcodine as Cough Suppressants by a New Method of Sequential Analysis', *Lancet* **1**, 860–862.
19. Thulbourne, T., and Young, M. H.: 1962, 'Prophylactic Penicillin and Postoperative Chest Infections', *Lancet* **2**, 907–909.
20. Weatherall, M.: 1982, 'An End to the Search for New Drugs?' *Nature* **301**, 387–390.

ANDRE de VRIES

COMMENT ON MICHAEL RUSE'S ESSAY

Professor Ruse deals with the question of the interruption of clinical trials when he refers to "sequential testing" by which a trial is "cut off" when a certain level of statistical significance has been reached. In essence, however, his discussion applies to a wider scope, that of the level of statistical significance demanded in clinical trials in general (not only sequential ones) including those where the endpoint-time is preset. Thus, although the title of his essay may seem too narrow for its content, the wider scope of the analysis makes it the richer.

Ruse's discussion of the requested statistical significance level in clinical trials is a timely one in view of the methodological uncertainties and the ethical dilemmas worrying all those responsible for the designation and carrying out of such tests. On the one hand, as Ruse points out, all involved wish the utmost safety of the test subjects; on the other, the demand of utmost safety is often a deterrent to the initiation or completion of the trial. Indeed, it is claimed by not a few health researchers, that the requirement of safety, as manifested in the demand for a "highly" statistical significance level of the test results, may and already has impeded the advance of medicine and, in some instances, harmed potential beneficiaries of new, as yet unreleased, therapeutic measures.

It is Ruse's contention that, although clinical trials "are essential for effectiveness and safety," and that "intuitively" he adheres to the demand of "very high levels of significance ($p<0.01$)," he rationally is prepared to cede to "lower," although still accepted, valid levels of significance "($p<0.05$)". It seems to me that this low, still "normal" level of significance, is more a manifestation of Ruse's methodological and ethical caution, than that, in analyzing the problem in question, it should be taken as an absolute lower limit. Indeed, and Ruse refers to it, statistical significance levels are arbitrary in themselves, and there may be and apparently have been cases in which clinical trials, in the classic sense, are or were not necessary at all. May I add that for some clinical trials, although originally claimed significant, their validity has in the course of time become doubtful, to wit, certain prolonged thrombosis-prevention trials.

In making the case for a reduction of clinical trial statistical significance levels, Ruse analyzes some motivations of the people involved in the trials, specifically the researchers, the subjects given drugs, the control subjects, subjects suffering from the disease to which the trial is relevant, physicians active in carrying out the trial, sponsors – whether publicly funded or privately financed (drug manufacture firms), and the general public. He then utilizes this analysis in his argument for the loosening of the statistical significance level stringency. Although, no doubt, some of these motivations have influenced the determination of "cut off" points or significance levels in given clinical trials, it seems to me that in the establishment of a reasonable future policy for the conducting of clinical trials one should exercise caution in attaching to these motivations too much importance. In fact, when taking such motivations into account as determining factors, one would derive the 'ought' from the 'is', or prescription from description. An ideal policy belongs to the 'ought' and not, or not necessarily, to the 'is'. Whether it will be possible to formulate an ideal clinical trial policy or whether this is an illusory undertaking, is another matter. The uncertainty, however, should not deter us from aspiring to the ideal.

The question then arises: What ought such an ideal policy be? Evidently, since particular clinical trials each deal with different situations, the ideal cannot be one general formula, but, as stated by Ruse, requested statistical significance levels ought to vary with the characteristics of the case, such as the severity of the disease, its spread, the knowledge about the drug to be tested, the drug's pre-human test performance, etc. Here one can only but agree with Ruse that "there are very good reasons why one should not stipulate criteria which are extremely high" and, "indeed, there may be times – many times – when it is reasonable to accept far-from-definitive evidence."

It seems to me that the importance of Ruse's analysis lies foremost in the stimulation of thinking about how much a clinical trial policy of varying stringency can be put into action, the practical step, which Ruse himself hardly touches. Evidently, the proof of the pudding lies in the eating. Who, or what body, will determine the statistical significance level for any specific clinical trial? Who will determine adequacy, morality, and cost containment? Will such procedures be workable? Satisfactory?

I am convinced that whoever will participate in clinical trial decision-making, the closest involved must be the test subjects and the physicians

– this in view of the evident consideration that, in the final issue, the aim of a clinical trial is the advance of therapy, which is the essence of the doctor/patient relationship.

It also appears to me that with the growing awareness of the patient's autonomy and his or her rights and responsibilities, the major role in clinical trial decisionmaking will eventually be assigned to the patient.

Finally, I hold that the inherent uncertainties in non-stringent clinical trials, and in clinical trials in general, ought to be and probably can be compensated by intensification of post-marketing drug surveys, an undertaking itself beset with considerable strategic difficulties – indeed an urgent theme in its own right.

New thinking about this complex problem of the practice of clinical testing – pre- and post-marketing – is urgently needed. Professor Ruse's reflections on the complex issues ingredient in the conducting of clinical trials with human subjects come as a welcome response in a time when we are urged to confront both theoretical and practical difficulties without the luxury of too much leisure time.

Tel-Aviv University
Tel-Aviv, Israel

SECTION IV

OBLIGATIONS AND THE AVOIDANCE OF INJURY

ARTHUR L. CAPLAN

IS THERE AN OBLIGATION TO PARTICIPATE IN BIOMEDICAL RESEARCH?*

While there is no dearth of writing about ethical issues regarding human experimentation in the literature of bioethics, surprisingly little has been written, particularly in recent years, about the question of whether or not any type of obligation might exist to participate in medical research. Most contemporary discussions focus on the adequacies and inadequacies of informed consent in combination with institutional peer review boards in protecting the welfare of subjects against abuse by researchers. Or, to put the point more accurately, the discussion of the ethics of research involving human beings currently consists of adversarial haggling about the details of specific legal and regulatory provisions concerning exactly what research must be reviewed, who is competent to consent, who ought to do the reviewing of experimental protocols, and exactly what degree of detail makes for a truly informed consent [20].

A powerful impression conveyed by all this attention to consent and committee review is that the main problem in the area of human experimentation is not assuring that an adequate number of subjects are available for biomedical research, but, rather, the prevention of the practice of murder or mayhem against those poor unfortunates who fall into the maw of biomedicine. While it is true that one does occasionally encounter in the current literature on the regulation of research the grousings of an odd researcher or two concerning the deleterious effects ethical concerns have supposedly had on the speed of the overall scientific research endeavor [18], the major preoccupation of most scholars who have written about the ethical aspects of research involving human subjects is protecting those unfortunates who wind up in research settings as a consequence of infirmity, insanity, or inanity against the malicious intentions of research scientists.

I. MORAL 'SCANDALS' AND THEIR PREVENTION

In noting the strongly protectionist character of extant literature on the ethics of human experimentation, I do not mean to suggest that those involved as subjects in research would be better off without the benefits of both informed consent and prior committee review. While there are many reasons for doubting the sufficiency of these modalities for protecting the interests of those who serve as subjects, there can be little doubt that the legal provisions now current in the United States[1] have done much to eliminate most of the egregious examples of moral atrocity from the domain of human experimentation. Nevertheless, the fact that the current system of regulations and protections is often declared a success on the grounds that the flagrant moral horrors of the past – e.g., the Tuskegee syphilis study, the Jewish Chronic Diseases Hospital case, the Willowbrook hepatitis studies, the Southwest Foundation placebo contraceptive trial – no longer occur says something quite telling about the true nature of the moral concerns which originally fueled the establishment of these protections.

A quick review of the articles and books written during the 1960's and early 1970's on the subject of the ethics of human experimentation reveals the importance of the role played by the description of concrete examples of subject abuse by specific researchers in bringing about regulatory action. The landmark studies of Henry K. Beecher, Paul Freund, Jay Katz [5] [6] [12] [16] and many others drew public and academic attention to the problem through the extensive use of case presentations of outrageous scandals or highly dubious practices in biomedical research settings. This tradition of motivating concern about the ethics of human experimentation on the grounds of the dangers of risk and harm continues today in both the anthologies and textbooks of bioethics [3]. It is easy enough to understand why so much effort has been put into the provision of the protections of informed consent and peer review when one realizes the historical role played by scandals in compelling public attention to the ethics of human experimentation.

The lessons drawn by various scholars [1] from the Nuremburg trials of those involved in barbarous medical experimentation during World War II focused ethical discussion on the protection of subjects to the exclusion of all other matters. At Nuremburg it quickly became evident to all what science and medicine could and would do in the service of the majority. Nazi medical experiments exemplified what crass utilitarian concerns for benefitting the majority within a society could lead to if followed faithfully and systematically.

The defense of Dr. Karl Brandt, one of the most notorious of the Nazi physicians involved in human experimentation, was built primarily around the moral legitimacy of a state's requiring sacrifices up to and including death from its members in the name of the general social good. Brandt's defense lawyers pointed out time and again during the course of Brandt's trial that the United States, France, and Britain had constantly demanded just such sacrifices from their citizens throughout the duration of the war by maintaining systems of conscription to insure adequate numbers of men in their respective armies. These states all believed that the clear risks to life and health resulting from war could be placed upon certain citizens in the name of the common good [1].

The need to protect individuals against the enormous power of unbridled utilitarianism – that is, majoritarian self-interest run amok – resulted in the post-war years in the promulgation of various codes of ethics which required informed consent in all areas of biomedical research [6]. Moreover, a large number of books and articles appeared which argued persuasively for the importance of obtaining informed consent from all competent subjects in order to assure that abuses would not arise in the conduct of research [5] [6] [12] [13] [16].

The moral imperatives of assuring the welfare of subjects against individual researcher abuse and the power of state-sanctioned utilitarianism resulted in the complex array of codes and regulations which now govern human experimentation in many nations. However, the provision of safeguards for those actually involved in human experimentation ought not permit moral complacency on the part of either researchers or public officials. For despite the existence of regulatory protections, the basic public policy issue of who should be asked to serve as subjects in biomedical research has not been adequately debated.

II. WHO ARE THE SUBJECTS OF BIOMEDICAL RESEARCH?

Unfortunately, very little is known about the composition and profile of the persons who currently participate in biomedical research in the United States (or in other nations). As a recent Presidential Commission has observed,

"[N]o data are collected from which to determine either the number of human subjects involved in Federally supported research in a given year or the nature and incidence of serious injuries associated with such research" ([20] p. 23]).

For the most part, the only data that are available are rough sketches of the kinds of research that is being conducted: therapeutic and non-therapeutic, invasive versus non-invasive, psychological and medical, along with some very rough measures of the sex, race, age, and social class of those who participate [9] [20]. These rough measures shed little light on the way in which subject participation is obtained, and, more importantly, on the specifics of how the benefits and burdens of biomedical research are distributed among the members of American society.

Yet, there are ample reasons for suspicion that inequities do exist today concerning the allocation of the benefits and burdens of biomedical experimentation. One study conducted by the sociologist Bernard Barber observed that,

> ... studies involving greater therapeutic benefit for the subjects are more likely than those of less benefit to be done using subjects the majority of whom are private patients, whereas studies with minor or no benefit for subjects are most likely to involve mostly ward or clinic patients ([4], p. 55).

The findings of Barber and his colleagues about ward and private patients in the United States are borne out by even the most casual observations of large medical research centers in many countries. The poor, the sick, and minorities seem to be utilized more frequently than other persons are in studies where little direct or immediate therapeutic benefit is expected. Physicians and medical students in most research hospitals will readily admit that they are much more hesitant to utilize wealthy or paying patients in research protocols than clinic or poor patients.

Even in nations where public and private systems of care do not exist, inequities still obtain in the participation of individuals in medical experimentation. As one Israeli writer has noted in a recent article,

> Most of the subjects for experimentation are ill people. There are two reasons for this phenomenon: the patients are within reach of the physicians, and, because of their special condition, they are inclined or even bound to submit to experimentation.
>
> Limiting the experiments to ill people only is undesirable, unfair and ineffective ([7], p. 4).

Even when matters of race, class, and religious affiliation are set aside, the fact remains that the bulk of experimentation is conducted on those members of the community who are sick, institutionalized, or poor. The number of healthy persons involved in medical research of their own free choice constitutes a miniscule proportion of the total number of individuals who now participate in experimentation.

III. BAD ETHICS AND BAD SCIENCE

The benefits enjoyed by the general public – improved medical care, better drugs, and increasingly powerful technological therapies – appear to have been acquired at a price that has been paid primarily by the sick, the poor, and the socially disadvantaged. If it is true that the rich, the healthy, and the middle classes do not do their full share in carrying the load of medical research, if they are behaving in the classical economic sense as "free riders," then it will not be sufficient to focus moral discussions concerning human experimentation solely on the adequacy of informed consent and peer review as vehicles for providing adequate protections to human subjects. While it is surely morally laudable that many societies have instituted various protections against both researcher and state abuse in the context of human experimentation, the moral value of such protections pales if it is the case that some groups are far more likely to need to avail themselves of these protections than are others.

It should also be noted that there are purely methodological reasons which strongly favor the broadest possible involvement of all segments of society in biomedical research. The biases introduced by the selective use of volunteers or sick populations in biomedical research are of great concern to those who conduct both medical and biobehavioral research [8] [22] [24]. The use of only certain classes of society – the institutionalized, the poor, college and medical students, the very ill – as the basis for claims concerning the efficacy of various medical treatments or interventions is highly suspect. Indeed, some researchers have noted that the under-representation of the elderly as participants in the testing of new drugs in both England and the United States is dangerous to the health of this population [24]. Similar problems have been noted with respect to the under-representation in research trials of women of child-bearing age, children, and fetuses. Not only are there reasons for concern about inequities in the participation of citizens in biomedical research on grounds of equity, there are sound methodological reasons for concern as well.

IV. THE MORAL BASIS FOR OBLIGATORY PARTICIPATION IN BIOMEDICAL EXPERIMENTATION

As noted earlier, there are few discussions in the literature of bioethics concerning the issue of the morality of participation in medical research.

When the issue has been addressed it has been couched in two ways:

(1) An obligation or duty to serve arises as a consequence of the benefits that will accrue to all from such participation, or,
(2) some type of social contract exists which places a duty on every citizen to participate in medical research.

Some physicians, such as Walsh McDermott, Louis Lasagna, and Leon Eisenberg, [10] have argued in favor of a duty to participate in research on the basis of the benefits that accrue to the public in the form of knowledge, safety and health. None of these physicians, for obvious historical reasons, grounds such a duty in the right of the state to claim participation in the name of the social good. Rather, the duty arises from the fact that they believe health, safety, and knowledge constitute what are often termed "public goods" – goods that accrue not to a majority of the members of a society, but to *all* members of society. The duty to participate in research is derived from the fact that the generation of these public goods requires public participation.

The counterarguments to this line of argument have been persuasively stated by, among others, Hans Jonas and Charles Fried [13] [15]. Both of these authors note that (a) it is not self-evident that health, safety, and knowledge are public goods since they are not always readily and freely available to everyone, (b) the "moral pull" exerted by the need to create public goods is outweighed by the far more powerful moral force of the need to respect human autonomy and dignity, and (c) that even if it is granted that safety, knowledge, and health are public goods, the object of research is melioristic – citizens have no duty to *advance* or *improve* public goods, particularly when it is not clear that any grave danger is posed for society by not doing so.

These arguments seem sound. While it may be true that the advancement of knowledge and the improvement of health care benefit many members of society, it is not evident that all persons are equally benefitted in the way that all persons benefit from paradigmatic examples of public goods, e.g., clean air or the existence of powerful armies for national defense. While the benefits of a strong army accrue equally and unavoidably to all, the same cannot be said of the benefits of health and well-being produced by biomedical research.

Moreover, many members of contemporary American society believe that knowledge, safety, and health are neither public nor good. There is sufficient skepticism in most societies about the value of medical interventions as to cast doubt on the validity of the claim that health or knowledge are public goods.

The other strategy for generating a duty to participate in research is to justify such a duty on the grounds of the existence of a "social contract." Since, the argument runs [15], we have benefitted from the participation of previous generations in biomedical research, we owe our own participation in research to future members of society, born and unborn, as a way of discharging this debt. Proponents of this thesis argue that while it may be true that no formal affirmation of consent to a social contract actually occurs, we can presume the existence of a tacit consent on the part of those who accept the benefits of knowledge and skill in the biomedical sciences. By accepting the benefits of scientific and medical knowledge in the form of therapy, improved diet, a higher quality of life, and so on, we affirm, obliquely but overtly, our duty to those individuals whose past sacrifices made these benefits possible, and incur a powerful moral obligation of reciprocity to bear additional burdens so that future generations may reap similar sorts of benefits.

But, it is not at all clear that those persons who participated in research in the past did so with the thought that they were creating a debt that had to be discharged by those who reaped the benefits of their service in experiments. Indeed, as Hans Jonas rightly observes [15], such a view diminishes the sacrifice of previous generations of research subjects by robbing their participation of its virtue. Many, such as Walter Reed and his research team, who participated in dangerous and deadly experiments, did so out of a desire to benefit humankind by their sacrifice. It is simply a conceptual mistake to think that a duty to reciprocate is the obligatory response to a gift which is freely given; at most, we are obliged to be grateful for a gift.

It would, of course, be naive to think that all those who participated in biomedical research in the past did so out of pure altruism. Many subjects were compensated for their participation; others were coerced or tricked into participating. However, it is hard to see how any obligation is generated among the existing members of a society who derive the benefits of research that was the product of compensation or duplicity. Furthermore, it is difficult to understand how obligations could exist to participate in research to any persons other than those who actually participated in past research which produced discoveries or benefits that have actually been utilized by current members of society. The only persons with a contractual claim on the living would be the members of past generations who gave of themselves with the thought that such giving had to be reciprocal. Unless these claims are cashed in, it seems mistaken to speak in a general way about duties to society or to past generations, and positively confused to invoke duties to participate in research on the basis of the claims of future generations of humankind. As was the case with the argument from the common good, it is hard to see how an argument from contractual reciprocity can be mounted which is sufficient for generating a duty to participate in biomedical research.

The arguments against the duty to serve as a subject in biomedical investigations seem to have been so persuasive as to have made the topic otiose. While there have been discussions in recent years of the moral acceptability of involving prisoners, children, the retarded, and other classes of so-called "vulnerable" subjects in research, these discussions have hinged upon the reliability of the informed consent mechanism for protecting these classes of persons from abuse rather than over the issue of an individual's duty to serve in research.[2] Nevertheless, the fact remains that inequities still exist in the ways in which the burdens and benefits of human experimentation in the biomedical sciences are allocated in North American and other societies. This fact demands that we seek other moral grounds in support of a duty to participate in research besides social utility, the need to serve the common good, and tacit consent to a mythological social contract.

V. CAN AN OBLIGATION TO PARTICIPATE BE GENERATED?

One way of generating a moral duty to participate in biomedical research is to acknowledge the fact that the common or collective status of the goods or benefits of research – knowledge, health, and safety – can be questioned, but to insist that those individuals who accept the benefits of such research nevertheless incur certain obligations, central among them a duty to participate in research. John Rawls has coined the term "fair play" to describe moral practices which prevail in certain forms of voluntary social cooperation. Rawls ([21], pp. 108–114), as well as H. L. A. Hart [14], have noted that those persons who benefit from participation in various cooperative social schemes – for example, the creation of a food co-op, a block patrol, a day care center, a mutual aid society, a rescue society – bear obligations to each other when called upon to bear the risks or burdens that involvement in cooperative endeavors often entails. In Hart's words,

> ... when a number of persons conduct any joint enterprise according to rules and thus restrict their liberty, those who have submitted to these restrictions when required have a right to a similar submission from those who have benefitted by their submission ([14], p. 70).

The members of a voluntary cooperative group can legitimately expect each group member to accept the burdens and risks of participation in such enterprises when the members have profited in some way from the activities of the group. The members of such cooperative associations are on sound moral footing in chastising and excluding any "free riders" that are discovered taking benefits but shirking responsibilities.

Rawls's notion of fair play, which properly governs the actions of the members of voluntary associations and organizations, has elicited a great deal of critical comment in recent years [2]. One central complaint is that his discussion fails to distinguish those who act as free riders in deriving benefits from various social enterprises, and those who simply have such benefits forced upon them without their consent.

For example, Robert Nozick has argued ([19], pp. 90–97), that no obligations can be said to be incurred by persons who inadvertently or unavoidably derive benefits from social schemes and cooperative endeavors which they neither approve of nor consent to. If I am lucky enough to benefit from the security afforded my neighborhood by the existence of a neighborhood security patrol, I nevertheless incur no obligation to pay for or serve on such a patrol merely because I chance to live in the locale where such a group is active. Nor do I incur any obligations or duties to the members of social groups merely because they choose to make me the beneficiary of their group largesse – if the Moonies bestow flowers upon me at the local airport, contrary to their hopes, I am under no obligation to read their literature or donate money to their organization.

Many philosophers have found Nozick's counter-objections to the notion of fair play persuasive. Certainly it seems odd to hold every citizen hostage to the benevolent intentions of group enterprises. Yet, as A. John Simmons [23] among others has pointed out, it is possible to salvage the principle of fair play as an independent mode of generating obligations and duties if its scope is restricted in certain ways.

As Simmons correctly observes, it would surely be ridiculous to argue that anyone who benefits from the activity of a social cooperative or group enterprise is thereby obligated to bear the burdens of that enterprise. For example, while a block patrol may afford me greater security in my own home or on my street, it is hard to see why I would be obligated to serve a tour on patrol simply because others have decided that they wish to increase their security in this way. Non-consenting bystanders simply do not incur obligations as a result of group activities regardless of how beneficial or desirable they may be.

However, as Simmons also observes [23], not all beneficiaries can be classified as non-consenting or innocent. If persons consciously seek out the benefits of a group enterprise or social endeavor; if, over time, they make no effort to avoid, whenever possible, the benefits of such activities; and if, over time, they continue to go along with a particular scheme or group activity when it would be a simple matter to avoid the benefits of that scheme, then what we have are not innocent beneficiaries but tacitly consenting "free riders" in the Hart and Rawlsian sense of the term. If people are thrusting books under my door as part of a neighborhood book club scheme and I make no effort to lock my door, post no notices to tell them to cease and desist, and if I eagerly read every volume that winds up in my house, then I do seem to be violating some sense of fair play if I refuse the book club's request to spend a few hours distributing books.

Even in the absence of overt or explicit consent, individuals do seem to be obligated to share in the burdens of group enterprises if they (a) consciously and knowingly accept the benefits of such enterprises, (b) make no effort to avoid receiving benefits, (c) fail to act on opportunities to avoid receiving benefits, and (d) consciously and knowingly seek out occasions in which benefits can be obtained from a social enterprise or scheme. Taking benefits from social enterprises does generate obligations of reciprocity in those circumstances where it would be a simple matter to avoid receiving such benefits.

If the concept of a group enterprise is broadly construed along the lines intended by Hart and Rawls, then it seems appropriate to describe the behavior of many of those who accept the benefits of modern medicine as falling under the rubric of fair play. Modern medicine is a vast social enterprise in which certain benefits are produced at the cost of various burdens, which include the need to conduct medical research.

If individuals consciously, knowingly, and continously accept the benefits of medical care by seeking out physicians and hospital personnel when they are ill, then they would seem to meet the conditions for being bound by the principle of fair play. If the only way the knowledge and skills utilized in modern medicine can be generated is through research involving human subjects, then those who accept the fruits of such research would appear to be under a duty to bear the burdens of research when called upon by the group to do so.

The principle of fair play can, I believe, be used to generate the moral foundation for broader public participation in biomedical research. Few patients are innocent bystanders, unwilling recipients of medical care. Most actively seek out the highest quality care they can possibly receive, and in so doing, cement their status as obligated participants in an ongoing cooperative enterprise.

VI. THE EXAMPLE OF THE TEACHING HOSPITAL AS A SOCIAL COOPERATIVE

An interesting case example of the way in which a health care institution functions as a social cooperative is the manner in which many teaching hospitals obtain patients for teaching purposes. Teaching hospitals often have "higher levels" of care available than do other institutions. Yet, the maintenance of this high level of care entails that certain burdens be shared by those who take advantage of the benefits that the teaching hospital has to offer.

Physicians often claim that patients who receive care in such institutions have an obligation to serve as subjects for teaching and educational purposes. If it is the case that patients do in fact understand the nature of the institution, I believe these physicians are correct. Those who choose to receive care in the context of a teaching hospital do incur an obligation to serve as the subjects for various teaching activities.

The teaching hospital is an excellent example of one of the types of social cooperative that exists in modern medicine today. Physicians, scientists, and patients organize themselves into a social unit to promote certain ends and to obtain certain benefits. Those who receive the benefits of such a scheme can legitimately be said to incur certain duties as a result. Fair play requires that those who knowingly and willingly choose to accept the benefits of better care, closer attention, and higher levels of provider competence incur a duty to serve as the subjects of medical teaching as long as the distribution of this burden and the benefits of better care are equitable and unbiased.

Of course it must be noted that there are some particular burdens that would not be reasonably assignable simply from a sense of fair play among participants in these kinds of hospital settings. No one is obligated to serve as a subject of the effects of lethal drugs on human beings. The fact that those who knowingly receive the benefits of care in a teaching hospital incur a duty to serve does not mean that such a duty immediately abrogates other rights they have, including the rights not to be harmed, killed, and to be fully and completely informed.

It is important to note that those who work in teaching hospitals, while believing that a duty exists on the part of those who are treated there to make themselves available to students and faculty for the purposes of teaching, usually realize that such a duty cannot be invoked casually or simply to suit the teacher's convenience. Patient consent is usually and should always be obtained from those asked to serve, both with respect to the specifics of an examination or demonstration and with respect to the convenience of the patient. The fact that a general duty to serve as a subject for teaching purposes exists among patients who choose to be in teaching hospitals does not mean that a person thereby abandons all other rights and entitlements in the face of such a duty, or that consent is not a necessary part of the discharge of any claims made upon such a duty.

VII. SOME PATIENTS HAVE A GENERAL OBLIGATION TO SERVE AS SUBJECTS IN BIOMEDICAL RESEARCH

An argument analogous to that regarding teaching institutions can be made about those who knowingly benefit from or receive care in institutions that openly and avowedly identify themselves as research institutions. Fair play requires that those who profit from the therapeutic benefits to be obtained as a result of the pursuit of greater knowledge and skills in such settings can be called upon to bear the burdens and costs coincident with such activities. As long as the patients in such settings freely and knowingly choose to receive their care in such settings, then they have a general obligation ([11], chap. 5; [14]) to participate in biomedical research.

There are, however some important restrictions on those who wish to involve persons as research subjects in such institutions. These include: (a) patients must be competent if they are to have any obligations under a principle of fair play, (b) they must understand that the institution in which they are to receive care is committed to research, (c) there must be a fair and equitable scheme for allocating the costs and burdens of research (it will not do to have only some individuals serve in all experiments), (d) no one is obligated to participate in experiments which pose significant risks to their health or well-being, (e) consent must be obtained from all individuals who are asked to participate in research, since choice is a prerequisite for invoking the norm of fair play in any social cooperative, (f) each subject must have a fair chance of obtaining the benefits derived from research, (g) no one is obligated to participate in research from which no benefits in terms of knowledge, therapy, or safety can reasonably be expected.

Many patients cannot meet some or all of these conditions. For example, those who are admitted for emergency care often are incompetent or unable to choose whether or where they will receive treatment. Others, due to age or mental impairment, cannot reasonably be said to have knowingly and freely chosen to receive care in a particular research setting. Still other persons may find it intolerable or merely inconvenient to serve in a research protocol at a particular time. A general obligation to serve as the subject of research in an institution that has research as one of its primary goals does not invalidate other rights that individuals have, including the right not to be harmed. Obviously, for some forms of research the right to control one's own body will come into conflict with the general duty to serve as a subject. But this is precisely the point – that oftentimes moral conflicts will and should arise in research settings when fair play clashes with personal safety, well-being, and autonomy.

VIII. THE LIMITS OF THE DUTY TO SERVE

If it is accurate to say that biomedical research and its constitutive institutions, organizations, and units constitute kinds of voluntary social cooperatives governed by the principle of fair play, then a number of interesting ethical consequences follow. First, biomedical researchers must recognize the distinctive nature of their enterprise and fully inform those persons who have an interest in reaping the therapeutic benefits of research about the choices they have and the burdens and costs they are likely to incur. Second, as members of such social cooperatives, biomedical scientists must also be considered eligible for bearing the costs and reaping the benefits of the enterprise – the duty to serve in experimentation is one that accrues both to patients and experimenters. Third, the state *qua* social institution is not licensed to make any demands on those parties involved in biomedical research other than those demands usually made on any voluntary social cooperatives – the activities must not interfere with or violate the rights of persons or groups who do not choose to associate with such enterprises, etc. The members of the research cooperative are the only persons licensed by the existence of the relevant cooperative activities to make demands on each other in the name of fair play. Finally, the existence of duties resulting from the principle of fair play that governs behavior within cooperative enterprises, such as biomedical research institutions, does not abrogate the obligations of researchers to respect the basic rights of patients who choose to be in such settings – the duty to participate in research is limited by the obligation to obtain informed consent and to have all research approved by institutional review boards.

Tensions will inevitably arise in determining when the duty of fair play has been satisfactorily discharged and on what occasion it is appropriate to discharge the duty. It may also be possible in some institutions to substitute alternative forms of service or even cash payments as methods of discharging the general duty to serve. Indeed, in many institutions patients are charged for their care and this fact must weigh heavily in determining the types of duties that patients can legitimately be said to possess if payment for care is required. Nonetheless, despite the thorny and complex issues involved in determining when and how the general duty may be discharged, the fact remains that, at least in some medical institutions, for those who knowingly and willingly seek to receive care they are, in so doing, consenting to the receipt of benefits that obligate them to participate in research.

IX. DO PATIENTS HAVE CHOICES?

If biomedical research institutions are instances of voluntary cooperatives akin to neighborhood security groups, social clubs, and other such institutions and organizations, the question arises as to whether anyone who derives benefits from the existence of research activities in such institutions can legitimately be said to be a knowing participant in the enterprise and thus the bearer of a duty to incur the cost involved in producing the benefits – in this case, the duty to participate in research. Even if the special cases of children, the comatose, the senile, the mentally ill, and the retarded are put aside, the question may still be raised as to whether even competent adults have any choice about whether they will or will not avail themselves of the services of a particular biomedical institution.

The answer to this question pivots around the response made to the proper description of the circumstances under which benefits are received. Is it the case that those who utilize research hospitals have no choice in the matter? Are they best described as the hapless and innocent beneficiaries of a social practice of which they do not approve and in which they have no desire to participate? Or, is it the case that those who reap the benefits of care in our most modern and advanced biomedical research institutions are more accurately described as actively and zealously pursuing the benefits of such research?

One criterion alluded to earlier for assessing the participatory status of patients, is whether they have made or would make any serious effort to avoid receiving health care benefits. Those who take benefits unwillingly or unwittingly have no duty to participate in research.

However, surely there are a large number of individuals who knowingly and freely choose to receive "state of the art" medical care. Indeed, many hospitals and research centers tacitly appeal to the choices such people make in selecting health care institutions by publicizing their reputations as "centers of excellence" or as "leaders in the field." While it is true that sick or disabled persons may appear to have little choice as to whether or not they will receive some form of care, it is surely false to say they have no choice about the matter. The fact that some people choose to have nothing to do with doctors, healers, or health professionals when afflicted with various ailments and diseases confirms the fact that the sick or disabled, while surely facing unfortunate choices, still have choices to make.

Obviously, the key issue in assessing the moral obligations that exist among the members of any social group is whether the participation of its members can truly be said to be the result of free and voluntary choice. There are sound reasons for suspecting that this may not always be the case in describing the presence of some persons in research settings.

Some persons simply do not understand that in most advanced societies they have a choice between receiving care in an institution that does not do research and one that does. Other people receive care in research settings against their will – they may be prisoners, or they may have no choice in that the only institution in their area is one that conducts research. Still, other persons are excluded from receiving care in certain institutions because of an inability to pay. For persons in these circumstances the norm of fair play simply does not obtain since, while such persons may receive the benefits of research in the forms of improved care, new therapies, or better levels of skill in the provision of medical services, they cannot be said to have voluntarily chosen to become a part of a cooperative activity. But for most people voluntary participation in research institutions is, or at least could be, the norm rather than the exception.

A key restriction on the scope of the principle of fair play is that those who are participants in a social enterprise do so as a result of their own free choice. This requirement leads to another set of conditions that must be met for the principle to have moral force on a particular person: (a) anyone in a research institution must have a choice as to whether and where they receive medical care, and (b) they must be given the opportunity of moving to a non-research institution or out of the medical system at any time prior to receiving care. If these conditions are met, then there is no reason for group members to tolerate refusals to participate in some form of research. Which is to say, that an obligation certainly exists among all competent, informed, and freely participating members of any social enterprise to recognize the duties their participation may entail and to take actions to discharge them.

It should be added that there are various ways of discharging such an obligation. It is entirely possible for the members of a social cooperative to decide to tolerate a certain percentage of free riders if the group wishes to do so. It is also the case that members of a social cooperative may decide to allow various alternative means to service in research as a way of discharging an obligation to the group, e.g., monetary payments, alternative types of service to the group, etc. The point is that medical research institutions which satisfy the conditions of voluntary social cooperatives are within their moral rights to require or not require the discharge of the obligations which are generated as a result of accepting the benefits of medical research.

X. THE ROLE OF THE STATE

It bears repeating that the issue of enforcing compliance among the members of any voluntary social cooperative must be left to the members of any such group. There is no role for the state or any state-licensed authority to play in demanding the discharge of obligations generated by the norm of fair play. The argument presented thus far in no way supports the claim that the state or its agents are morally enfranchised to conscript or otherwise coerce persons into participation in medical research. At best, it is within the power of the voluntary cooperative to enforce the effective discharge of duties incurred on the part of those who accept the benefits of such a group. The only morally acceptable means of achieving this end is for the cooperative to simply exclude those persons who knowingly accept the benefits of a cooperative activity, but renege on their obligation to bear the costs associated with producing the benefits. Medical institutions which clearly and forthrightly identify themselves to patients as research institutions would be within their rights to exclude persons who refuse to participate in any form of research if they wish to do so. However, the state should have no jurisdiction over the enforcement of obligations and duties of participation in voluntary social cooperatives.

I have argued that (a) it does not make sense to view the benefits of medical research as public goods, (b) that it is possible to describe a set of conditions that must obtain if participation in medical research is to be described as informed and voluntary, and (c) that when such conditions are met, it is reasonable to speak of a duty to participate in research as a result of the principle of fair play which obtains in any voluntary social cooperative. There do seem to be a number of institutions involved in biomedical research in a variety of countries that do, in fact, meet these conditions, primarily so-called tertiary care facilities or teaching hospitals, and, thus, there do seem to be a number of biomedical research institutions which can rightly demand the participation of individuals in research in return for the benefits they will receive by choosing to receive care in such institutions, should the institutions choose to make such demands.

If my argument is valid, will the norm of fair play serve as a basis for redressing the present inequalities in the distribution of the burdens of experimentation current in many countries? What, if any, are the obligations incurred by healthy persons who have no ongoing involvement with any particular medical research institution? For the argument could be made that in a general way all citizens benefit from the existence of organized biomedical research in a particular state in that they enjoy the benefits of better public health measures, safer foods and medicines, and the security of knowing that competent medical care is available to them should they require it.

I doubt that the mere acceptance of safer foods or cleaner water supplies would suffice to obligate members of the general public to participate in biomedical research. In most cases, those who take these benefits do so either in ignorance of the fact that they are produced, at least in part, of biomedical research, or, because they have contracted for them by purchasing them in the open market. While it is difficult to see how patients who find their way to hospitals or clinics for care could reasonably be said to be innocent and unconsenting recipients of medical benefits, it is not so difficult to see how members of the general public might be the unwitting or unwilling recipients of the fruits of biomedical research.

It should also be noted, however, that at least some persons fall into a rather large grey area between those who actively seek out the benefits of biomedical research and those who are its innocent or paying beneficiaries. These are individuals who receive care from their own physicians or who receive care in institutions that do no biomedical research whatsoever. It is difficult to know whether such persons and their caregivers ought to be described as falling inside or outside of the social cooperative or enterprise of biomedical research. If physicians and patients in such settings do consciously strive to keep up with the latest research findings, to take advantage of the most current therapeutic and pharmaceutical knowledge, then it becomes more difficult to justify their exclusion from the general enterprise of biomedical research.

In part, the question of who is to be viewed as bound by the principle of fair play, and who is not, will turn on the matter of choice. But even if the scope of the principle is restricted in the realm of biomedical research to those institutions and their clients who consciously, knowingly, and continuously seek out the therapeutic benefits biomedical research has to offer, there would still seem to be a good deal of injustice and inequity in the current distribution of the research burden in contemporary biomedical research. The general obligation to participate in research in our large tertiary care hospitals and research institutes binds many more patients than are currently participating in or being asked to participate in biomedical research. The question of how and when this general obligation may be discharged is surely open to further discussion and debate. If the arguments presented here are sound, then there are valid moral grounds for requiring the participation of a broader (and more economically varied) segment of the population in biomedical research then is currently being utilized.*

The University of Minnesota
Minneapolis, Minnesota,
U.S.A.

NOTES

* I would like to thank Janet Caplan, Hans Jonas, Ronald Bayer, John Arras, and Anthony E. Gallo, Jr., for their helpful comments on an earlier draft of this essay. I would also like to thank the members of the Hastings Center staff and the philosophy department at Union College for their comments and suggestions when this paper was presented as an invited address. Finally, I would like to acknowledge the support of the Charles C. Culpeper Foundation in preparing this manuscript.
[1] See 45 CFR 46, 'Protection of Human Subjects', revised March 8, 1983.
[2] An interesting exception to this general trend is Richard McCormick's 'Proxy Consent in the Experimental Situation', *Perspectives in Biology and Medicine* **18** (Autumn 1974), 2–20. McCormick argues that in relatively risk-free situations parents can give proxy consent for their children to participate in non-therapeutic research on the grounds that parents have an obligation to cultivate sociality and virtuous conduct in their children.

BIBLIOGRAPHY

1. Alexander, L.: 1949, 'Medical Science Under Dictatorship', *New England Journal of Medicine* **241**, 39–47.
2. Arneson, R. J.: 1982, 'The Principle of Fairness and Free-Rider Problems', *Ethics* **92**, 616-633.
3. Arras, J. and Hunt, R. (eds.): 1983, *Ethical Issues in Modern Medicine* (2nd ed.), Mayfield Press, Palo Alto, California.
4. Barber, B. *et al.*: 1973, *Research on Human Subjects,* Russell Sage, New York.
5. Beecher, H. K.: 1966, 'Ethics and Clinical Research', *New England Journal of Medicine* **274**, 1354–1360.
6. Beecher, H. K.: 1970, *Research and the Individual,* Appendix A, Little Brown and Co., Boston, Massachusetts.
7. Carmi, A. (ed): 1979, 'The Challenge of Experimentation', in *Medical Experimentation,* Turtledove Press, Ramat-Gan, Israel.
8. Chalmers, T.C., *et al.*: 'Controlled Studies in Clinical Cancer Research', *New England Journal of Medicine* **287**, 75-78.

9. Cooke, R. A.: 1980, 'Some Notes on the Subjects of Biomedical and Behavioral Research', *Report prepared for the President's Commission for the Study of Ethical Problems in Medicine and Biomedical and Behavioral Research,* U.S. Government Printing Office, Washington, D.C.
10. Eisenberg, L.: 1977, 'The Social Imperatives of Medical Research', *Science* **198**, 1105–1110.
11. Fishkin, J.: 1982, *The Limits of Obligation,* Yale University Press, New Haven, Connecticut.
12. Freund, P. A. (ed.): 1970, *Experimentation with Human Subjects,* George Braziller, Inc., New York.
13. Fried, C.: 1974, *Medical Experimentation,* Elsevier, New York.
14. Hart, H. L. A.: 1955, 'Are There Any Natural Rights?', *Philosophical Review* **64**, reprinted in A. Melden (ed), *Human Rights* (1977), Wadsworth Publishing Co., Belmont, California, pp. 61-75.
15. Jonas, H.: 1970, 'Philosophical Reflections on Experimenting with Human Subjects', in P. Freund (ed.), *Experimentation with Human Subjects,* George Braziller, New York.
16. Katz, J.: 1972, *Experimentation with Human Beings,* Russell Sage Foundation, New York.
17. McCormick, R.: 1974, 'Proxy Consent in the Experimental Situation', *Perspectives in Biology and Medicine* **18**, 2–20.
18. Muggia, F. M.: 1981, 'Anticancer Drug Development and Federal Regulation: Protection Against Progress?', *American Journal of Medicine* **71**, 341–344.
19. Nozick, R.: 1974, *Anarchy, State and Utopia,* Basic Books, New York.
20. President's Commission for the Study of Ethical Problems in Medicine and Biomedical and Behavioral Research: 1983, *Implementing Human Research Regulations,* U.S. Government Printing Office, Washington, D.C.
21. Rawls, J.: 1971, *A Theory of Justice,* Harvard University Press, Cambridge, Massachusetts.
22. Rosenthal, R. and Rosnow, R. L.: 1975, *The Volunteer Subject,* John Wiley and Sons, New York.
23. Simmons, A. J.: 1979, *Moral Principles and Political Obligations,* Princeton University Press, Princeton, New Jersey.
24. Smith, C., *et al.*: 1983, 'Drug Trials, the Elderly, and the Very Aged', *Lancet* **266**, 1139.

T. FORCHT DAGI AND LINDA RABINOWITZ DAGI

PHYSICIANS EXPERIMENTING ON THEMSELVES: SOME ETHICAL AND PHILOSOPHICAL CONSIDERATIONS

I. INTRODUCTION

Physicians are generally taught that heroics are foolhardy and deviate from the standard of wisdom and reasoned judgment to which they are held. Nonetheless, there is a strong tradition of romantic heroism in medicine. For the most part, this tradition has focused on the image of the brave physician staving off the angel of death with a spear, even in the most hopeless of situations. This tradition encouraged dedication and selflessness, and promulgated the vision of the doctor as a stalwart and tireless protector of the patient's welfare. This tradition also embodied a sense of pride and of adventure, of disregard for the physician's personal safety, and of a willing self-sacrifice to the ideals of medicine.

Objectively speaking, the actions a physician would undertake in investigating an epidemic, in exposing himself to contagion, or in working himself to exhaustion would appear to contradict the dictum that heroics are foolhardy. It turns out that wisdom and professional moderation constitute only two of a number of competing values in the self-image of physicians. At the conjunction of brave, if foolhardy, dedication and careful scientific endeavor there stands a tradition of autoexperimentation in medicine. Autoexperimentation has been widely practiced but only rarely discussed. Altman has published a review and a recent book on the subject [1, 2] and Bean included some insightful observation on autoexperimentation in his work on medical experimentation in general [3, 4].

There are two aspects to what has come to be called autoexperimentation. One aspect is auto-observation: like Sanctorius, who weighed himself daily for decades to learn about nutrition and metabolism, the researcher will keep careful records of what happens to his body under circumstances which are more or less routine, even if controlled. The other aspect is autoexperimentation in a more active sense: like Forsmann, who subjected himself to cardiac catheterisation to determine its safety and efficacy, [7] or like Pettenkofer and Emmerich, who swallowed cultures of *Vibrio Cholerae* to show, misguidedly, that Koch was wrong about the etiology of cholera, [9] the researcher will subject himself to dangerous or potentially hazardous maneuvers in order to observe their effects on his own body. There is also some overlap between the two paradigms, as when an investigator subjects himself to changes in his routine which are known to be innocuous, but whose effects are unstudied, or when an investigator takes advantage of a famine, or a similar catastrophe or unusual occurrence he cannot control, and observes the consequences of limiting the measures he might otherwise have taken to protect his health.

Beecher commented in 1959 that "Experimentation upon other men requires a willingness to experiment on oneself as evidence of good faith, although in a given case self-experimentation may be wholly impractical" [5]: As Altman points out [1], by 1970, Beecher had substituted the word *implies* for *requires* in this quotation [6].

Autoexperimentation, at least in its more hazardous aspects, has been regarded with continuing ambivalence. Whereas the experiments themselves are considered foolhardy, and the data arising from them are usually criticised for their lack of statistical significance, the experimenters usually come to be regarded with respect and admiration, particularly if they prove that a procedure that had been considered highly perilous could be performed with an acceptable margin of safety. Insofar as success converts the investigator into a hero, failure can convert him into a martyr. In the romantic paradigm of medicine, neither eventuality is altogether objectionable. Thus, regardless of the scientific objections that accrue the practice of experimenting on oneself, the personal characteristics that become attributed to the adventuresome investigator have rendered this form of medical research attractive in many ways.

Altman lists seven reasons why autoexperimentation is pursued [1, 2]. Experimentation on oneself is convenient and reliable; it satisfies one's innate curiosity; it provides a known control; it allows a better comprehension of complications arising in the course of the study; it can benefit the investigator personally (as when he subjects himself to an untested, but promising treatment or preventative measure for an acquired or potentially acquired condition); and it simplifies the legal problems of obtaining experimental consent from potentially misinformed subjects. There are a number of other philosophical and ethical concerns which touch on this tradition, however, and it is the purpose of this essay to review and discuss them.

II. THE VALUES IMPLICIT IN AUTOEXPERIMENTATION

Since the Scientific Revolution, the physician – even the practicing physician – has been led to value clinical and basic experimentation both in a search for knowledge for its own sake, and for the acquisition of knowledge to relieve suffering and prevent disease. The physician will often have two competing, and sometimes conflicting roles impressed upon him: he is expected to be at once the physician, and simultaneously, the scientist. Whenever the physician acts in such a manner that the welfare of his particular patient is not his first and foremost priority, and the collection of data is, the physician strays from the role of the healer to the role of the investigator. Both roles are valued historically, and both are considered valid roles for the physician. Autoexperimentation is attractive because the physician steps out of the healing role and enters the investigative role to the potential detriment of no one but himself. No contract to heal is abrogated.

The romantic notion of the physician as a hero has already been referred to. It may be overstating matters to claim that most physicians enjoy some sense of heroic fantasy in their involvement with medicine, but, on the whole, saving lives is a very satisfying occupation. There is, therefore, an implicit value in dedicating one's life to the benefit of others, and this value extrapolated could lead one even to the point of imperiling one's own existence for these very same ends.

The ethos of medicine places a great value on the physician as the protector of his patients, as an extension, perhaps, of the approval gained by the man who risks his own life to protect another's. The physician who risks his own health not for the ideal of science, but for the specific benefit of an individual patient is therefore particularly celebrated, and may well become a heroic figure.

Western society is ambivalent in its attitude to suffering. On the one hand, so much of the resources of society have been dedicated to the improvement of man's lot that one could not possibly conceive of suffering evoking anything other than protest and resistance. On the other hand, many religious doctrines hold amongst their tenets a conviction that suffering improves the spiritual lot of man, and, far from deserving disparagement, suffering should be savored philosophically and appreciated morally. Thus the physician who endures suffering in the course of experimenting on himself exploits society's ambivalence towards suffering to the full. On the one hand, he demonstrates by risking his own welfare his dedication to the health of others, while on the other hand, he is morally probate and philosophically improved. Both of society's attitudes are thereby reconciled, and both sets of values are satisfied.

An offshoot of religious attitudes towards suffering is a value in medicine derived from the various formulations of the Golden Rule. One part of the physician's virtue devolves from his willingness to subject himself to his patient's suffering, both in terms of exposing himself to the risk of contracting their disease, and in terms of empathizing with and risking the discomforts of their treatment. The physician's virtue, in fact, is attested to both on account of his *personal* traits, and on account of his *professional* characteristics. By experimenting on himself, the physician satisfies both the areteic, or moral, and the non-areteic, or non-moral values implicit in the notion of a "good" physician. Thus, part of the attractiveness of autoexperimentation as a medical tradition is the opportunity it lends the physician to enact his appreciation of some very powerful values in medicine and society.

III. THE SCIENTIFIC AND LEGAL ADVANTAGES OF AUTOEXPERIMENTATION

The greatest pitfall in science is the random error or the uncontrolled event. Thus, it is argued, the more control the scientist can exert over his experimental subjects, the more likely he is to avoid this difficulty. In making a series of simple observations, for example, even if no invasive maneuvers are required, and even if the subject serves primarily as a control, strict adherence to routine is required. A casual subject might object to having himself weighed daily, or to having venipunctures at fixed intervals, and may abandon the discipline required for the experiment. A casual subject may leave the experiment, or may lose interest for various reasons.

The human subject must be advised fully of his risk and rights in the experimental setting. Consent becomes a legal as well as a philosophical issue, and the scientist might fear that truly informed consent with true appreciation of the risks is impossible in his particular experimental setting. Experimenting on one's self obviates this difficulty.

The dedicated scientist making observations on himself may recognize subtle changes in himself that a less dedicated, less expert, less attentive, and less curious subject might overlook. Thus, insofar as his data reflect his subjects' report, the physician's best data could conceivably come from observations on himself. He can continue to observe himself for delayed effects long after the formal span of the experiment has ended. The convenience of having the experimental subject always at hand cannot be overstated: the entire issue of trusting one's subject to report observations with consumate accuracy becomes obviated.

The main advantages of autoexperimentation, therefore, revolve around issues of control, accuracy, perseverence, convenience, ethics and consent. Relative to each of these concerns, it can be argued, the physician is better off observing himself, and drafting himself to be the subject.

IV. DIFFICULTIES ARISING IN AUTOEXPERIMENTATION

Despite the advantages of controlling the experimental setting as completely as the physician putatively can when he experiments on himself, the legitimacy of autoexperimentation is open to question on both philosophical and scientific grounds. Jakobovits, for example, argues that the Talmud prohibits autoexperimentation because of the prohibition on imperiling one's life, and on self-mutilation [10]. Both of these are, in W. D. Ross's sense [12], pp. 19ff), *prima facie* obligations that carry enormous weight, but that fall short of being absolute moral imperative. Thus, at least in Jewish law, there are times when the obligation to avoid lethal hazard and the obligation to avoid mutilation may be suspended. The obligations are not diluted in any absolute sense: simply put, other obligations of a more pressing nature (e.g., saving the life of another) can modify the force of the imperative. On the other hand, one is taught that "any chance to save life must be pursued at all costs," and "the obligation to save a person from any hazard to his life or health devolves on anyone able to do so" [10]. While no one has the right to volunteer his life, (i.e, one may not ordinarily sacrifice one life to save another, despite rare situations where martyrdom is required), "measures involving some immediate risks of life may be taken in attempts to prevent certain death later" [10].

Jakobovits derives the following conclusions from the principles above:

1. Possibly hazardous experiments may be performed on humans only if they may be potentially helpful to the subject himself, however remote the chances of success are.
2. It is obligatory to apply to terminal patients even untried or uncertain

cures in an attempt to ward off certain death later, if no safe treatment is available.
3. In all other cases it is as wrong to volunteer for such experiments as it is unethical to submit persons to them, whether with or without their consent, and whether they are normal people, criminals, prisoners, cripples, idiots, or patients on their deathbed.
4. If the experiment involves no hazard to life or health, the obligation to volunteer for it devolves on anyone who may thereby help to promote the health interest of others . . . [10].

According to Jakobovits, therefore, Jewish law recognizes a distinction between innocuous and hazardous experimentation. It also recognizes a distinction between information required for the benefit of a specific patient or a specific time, and experimental investigation of a more general sort, where the benefit of a specific patient with a particular diagnosis is not necessarily a major concern. The situation involving even highly experimental therapy for an otherwise lethal condition is considered in the first category when it is the patient with the lethal condition who is subject to the experimental (and even hazardous) therapy. The needs of a *specific* patient considerably strengthen the arguments favoring the suspension of other *prima facie* obligations in most circumstances. Experiments involving some measure of hazard to the subject are sometimes permitted when a potentially lifesaving or lifepreserving benefit may accrue a *specific* individual. Participation in innocuous experiments may well be *required* under the same or similar conditions.

The Nuremberg Code makes the following mention of autoexperimentation in Article 5: "No experiment should be conducted where there is an *a priori* reason to believe that death or disabling injury will occur; except, perhaps, in those experiments where the experimental physicians also serve as subjects" [13, 1]. One is given to believe that autoexperimentation is permitted in this formulation specifically to encourage the participation of physicians as subjects in their own experiments to serve as a procedural safeguard, a sort of procedural justice. The scientific validity of such experiments is not at issue, and neither does the Code address the case of physicians experimenting on themselves where only innocuous procedures are involved.

The Code for Self-Experimentation at the National Institutes of Health was intended to provide the same safeguards for physician-subjects as for other subjects [1]. It requires physicians to undergo a medical evaluation before beginning their protocol. The purpose of the medical examination is to assure the health of the investigator rather than the validity of the experiment: whether the investigator is, paradigmatically, an appropriate "normal" control or experimental subject is not the point.

It is pertinent, at this point, to review the meaning of the term "experiment." Altman is satisfied to use McCance's definition, which he calls "the broadest and perhaps the most realistic:"

> Let us start with the word experiment, which most biologists use very loosely to cover any investigation, however trifling, made to advance knowledge. The term generally implies some deliberate change of conditions without foreknowledge of the results but with subsequent observation of them. It may be used, however, even when the conditions are not being deliberately changed, when the term observation would be more correct . . . We should, I think, for present purposes, regard anything done to a patient, which is not generally accepted as being for his direct therapeutic benefit or as contributing to the diagnosis of his disease as constituting an experiment, and falling, therefore, within the scope of the term experimental medicine[11].

McCance's definition is sufficiently broad to include almost every nuance of experimentation on humans, but it begs some important distinctions. We propose to differentiate among several categories of human experimentation arranged in order of increasing risk:

1. simple, prolonged observation;
2. simple observation following the deliberate change of conditions known, or strongly believed to be innocuous;
3. the application of measures known or believed to be effective or innocuous designed to treat or to modify artificially induced, but innocuous conditions;
4. observation following the deliberate change of conditions without foreknowledge of their consequences;
5. the application of measures of uncertain safety or efficacy designed to treat or to modify naturally occurring conditions for which no satisfactory therapy exists;
6. the application of measures of uncertain safety or efficacy designed to treat or to modify artificially induced, but innocuous conditions;
7. the application of measures known to be safe and effective in treating or modifying artificially induced, hazardous conditions;
8. the application of measures of uncertain safety or efficacy designed

to treat or to modify naturally occurring conditions for which satisfactory therapy exists;
9. observation following the deliberate change of conditions with some expectation of hazard;
10. the application of measures of uncertain safety or efficacy designed to treat or to modify artificially induced conditions of uncertain safety;
11. the application of measures of uncertain safety or efficacy to treat or to modify artificially induced conditions of clearcut and definable hazard.

Two types of argument can be proposed concerning the legitimacy of auto-experimentation in relation to levels of risk: the first, derived from principles of beneficence, maintains that the lower the risk, the more legitimate the experiment utilizing human subjects, irrespective of whether the subject be the experimenter or some other volunteer. The second, derived from values of autonomy, courage, self-sacrifice, and empathy, and based in part on forms and interpretations of the Golden Rule, maintains that if it is *ever* necessary to perform human experiments of great risk, it may be not only permitted, but even required that the experimenter be willing to subject himself to the same hazards he would have others undergo.

In McCance's definition of human experimentation, procedures of an experimental nature that could, conceivably, benefit the patient directly are not specifically mentioned. We would divide each of our eleven categories into two groups depending on whether or not a direct benefit to the subject could be envisioned. The negative value of risk diminishes (i.e., most subjects would be willing to assume a greater risk) when the promise of greater benefit can be held forth. This observation holds true even when the physician experiments on himself: the degree of risk is a salient consideration.

V. CONSIDERATIONS OF LOW RISK

The objections to autoexperimentation in low risk situations reflect considerations beyond the potential hazard to the physical or psychological welfare of the physician-subject. Whereas the question of risk is not necessarily eliminated altogether from the discussion, it does not constitute the limiting factor.

Insofar as an autoexperiment utilizes only one individual as its study population, important scientific objections can be raised. One swallow does not a summer make, nor one subject a population. Nonetheless, if a physician takes other measures to assure statistical validation of his data, and utilizes himself as a pilot or a test subject, for example, or to obtain uniquely subjective insights into the effects of this protocol or that, the statistical objection dwindles in importance.

By the same token, the objectivity of data obtained through autoexperimentation is open to criticism. As with the statistical objection, however, this criticism can be resolved by experimental design. If the *purpose* of the experiment is to obtain some uniquely subjective insights (e.g., Freud's flirtation with cocaine, or Purkinje's experiments with foxglove leaf) then very little objection can be mustered so long as it can be shown that the subject's response is typical, and his observations are not adversely affected by the conditions of the experiment.

There is a unique but unfortunate situation that devolves when the physician falls ill and is in a position to subject himself to therapeutic experiments which he does not control. What can transpire in this setting is a combination of human medical experimentation of the usual sort and of autoexperimentation. The physician observes himself while being studied by others. One might query whether the very fact of being a physician, or of observing oneself being studied changes the validity of the experimental protocol or detracts from it, much as an anthropologist might find it difficult to study a primitive society without changing its routine *pari passu*. These important questions are outside the scope of this paper.

VI. HAZARDOUS EXPERIMENTS

The principle of autonomy underlies the contention that individuals may risk their lives when they choose to do so. Certain social, religious, or philosophical considerations may limit the extent to which an individual would exert his autonomy. All things being equal, however, there is no absolute objection to a physician's placing himself at risk, whether in scuba diving, mountain climbing, motorcycle racing, parachuting, or experimenting on himself. Under some circumstances, disregard for one's personal safety is celebrated as courage and bravery. Under other circumstances, disregard for one's safety is dismissed as foolhardy. Two major factors sway our judgement: the stakes, and the risk.

There is no good way to evaluate risk. When the risk is overwhelming, the stakes determine how an act is viewed. The ultimate and final risk is often (but not always) death. Dishonor, mutilation, ridicule, failure and exile are examples of risks which may be greater than the risk of death, however, in some settings. Authorized self-sacrifice is deemed martyrdom and is highly valued in almost all societies. Is there, then, an obligation on the physician to martyr himself through medicine, and expose himself to risk?

There is a difference between necessary risk and unnecessary risk. There are times, during an epidemic, for example, when it is incumbent on the physician to assume deadly peril in the course of his duties. This risk "comes with the territory." It is unavoidable. On the other hand, much as a policeman is permitted (and at times even required) to wear a bulletproof vest, the physician is certainly not enjoined from enlisting whatever protective measures he can muster. The physician is not required to put himself at any greater risk than his duties require. In epidemic situations, moreover, it has been argued that physicians should be immunized first because of their increased exposure and because of their social utility. On the other hand, when a physician encounters a condition he does not know, he may wish to study that condition despite the fact that he puts himself at greater risk, or he may wish to test protective measures on himself before exposing others to the risk that the protective measures might fail. In the United States Air Force, for example, there is a long tradition of flight surgeons testing life support equipment on themselves before approving the equipment for others; several flight surgeons died before techniques for parachuting from the upper reaches of the atmosphere were perfected [8].

No one can contest the courageousness of these autoexperiments, but they cannot be considered obligatory. They may satisfy an urge for adventure, or a certain sporting instinct that occurs in physicians as readily as in any other group in society. On the other hand, actions admired in the past are not necessarily required or even desirable in the future. Just because the early radiologists were unwitting victims of their own ignorance, and became the unfortunate population from whom the effects of radiation were learned is no reason why modern physicians should repeat their errors. The pioneers can be admired for their fortitude, and their characters can be idealized without their *acts* serving as models for the present.

VII. CONCLUSION

Discrete situations arise in which physicians are victims or observers of experiments of nature. In those settings, physicians may have a definite obligation to observe themselves and to record their fate. On the other hand, society has an enormous investment in the training of physicians and scientists, and the voluntary assumption of experimental hazards must be balanced against the responsibility invoked by this investment.

Physicians can make important observations on themselves. Before engaging in autoexperimentation, it is wise to verify the validity of these observations and to assure their objectivity. An advantage of autoexperimentation is the elimination of problems of informed consent. On the other hand, to act in accordance with principles of autonomy, the physician must satisfy himself that he is not deluding himself in considering himself psychologically and physically invulnerable, and that he is not denying himself the same intellectual honesty he would expect from another in consenting to be a subject.

There is no requirement to place oneself in jeopardy beyond the ordinary, if innumerable, risks attendant on the practice of medicine. Most experiments on one subject alone can contribute very little to our understanding of science and of medicine at this stage of our scientific development. Nonetheless, there are circumstances in which autoexperimentation will result in unique observations, and under those circumstances, with proper safeguards, the exercise of autoexperimentation can result in the execution of some valuable, and, betimes, even heroic acts.

Walter Reed Army Medical Center
Washington, D.C., U.S.A.

and

Georgetown University Hospital
Washington, D.C., U.S.A.

BIBLIOGRAPHY

1. Altman, L. K.: 1972, 'Autoexperimentation', *New England Journal of Medicine* **286**, 346.

2. Altman, L. K.: *1986, Who Goes First? The Story of Self-Experimentation in Medicine*, Random House, New York.
3. Bean, W. B.: 1952, 'A Testament of Duty: Some Strictures on the Moral Responsibilities in Clinical Research', *Journal of Laboratory Clinical Medicine* **39**, 3–9.
4. Bean, W. B.: 1977, 'Walter Reed and the Ordeal of Human Experiments', *Bulletin of the History of Medicine* **51**, 75–92.
5. Beecher, H. K.: 1959, 'Experimentation in Man', *Journal of the American Medical Association* **169**, 461–478.
6. Beecher, H. K.: 1970, *Research and the Individual*, Little Brown and Co., Boston, Massachusetts.
7. Forsmann, W.: 1929, 'Die Sondierung des rechten Herzens', *Klinische Wochenschrift* **8**, 2085–2087.
8. Hitchcock, F. A.: 1971, 'Paul Bert and the Beginning of Aviation Medicine', *Aerospace Medicine* **42**, 1101–1107.
9. Howard-Jones, N.: 1973, 'Gelsenkirchen Typhoid Epidemic of 1901: Robert Koch, and the Dead Hand of Max von Pettenkofer', *British Medical Journal* **1**, 103–105.
10. Jakobovits, I.: 1966, 'Medical Experimentation of Humans in Jewish Law', *Proceedings of the Association of Orthodox Jewish Scientists* **1**, 1–7.
11. McCance, E. A.: 1951, 'The Practice of Experimental Medicine', *Proceedings of the Royal Society of Medicine* **44**, 189–194.
12. Ross, W. D.: 1930, *The Right and the Good*, Oxford University Press, Oxford, England.
13. *United States, Adjutant General's Department. Trials of War Criminals before Nuremberg Military Tribunals under Control Council Law*, No. 10, October 1946–April 1949: The Medical Case, Vol. 2, 1949, U.S. Government Printing Office, Washington D.C., 181–183.

IRVING LADIMER

PROTECTION OF HUMAN SUBJECTS: REMEDIES FOR INJURY

Protection of human subjects is imperative for biomedical research and development to proceed effectively and ethically. Investigation and production of new drugs, devices, and methods are advancing rapidly throughout the world. In the United States, which has been notably slow, movement is evident. Relatively more drugs have been approved recently and review procedures are being improved. "The Food and Drug Administration approved 27 'totally new' drugs last year (1981) – the most . . . since passage of the 1962 legislation governing drug approval . . . more than twice the 12 approved in 1980" [8]. The drugs were indicated for a wide array of physical and mental problems and for diagnosis as well as therapy.

This heightened activity obviously implies more use of human beings, normal and sick, as subjects in varied research contexts and settings. Drug development for international markets is now significant in China, India, and third world countries. In the United States, the agency concerned with ethical issues in research, the President's Commission for the Study of Ethical Problems in Medicine and Biomedical and Behavioral Research, sought data on the extent of research involving human subjects. "Annual expenditures (in the U.S.) for health-related research are now about $8 billion for which the Federal Government contributes more than 60%; three quarters . . . from the Department of Health and Human Services, primarily through the National Institutes of Health" ([15], p. 12). A study conducted by a Task Force of the Department of Health, Education and Welfare (predecessor to HHS) in 1975–76, on compensating injured research subjects, used a figure of 600,000 for estimating costs, as the number of subjects in NIH-supported clinical trials [7]. This number excluded other HEW and Federal programs, all private drug and device firms and universities and research agencies funded from non-NIH sources. By any measure or estimate, the number of non-therapeutic and, particularly, therapeutic subjects must be substantial and clearly a matter of public concern and interest.

NUMBER OF INJURIES

More to the point but even more conjectural were the number and types of injuries sustained by subjects. In 1975, the Government conducted a sample survey, by telephone, to estimate the number and type of research-related injuries. The sample included 331 researchers studying nearly 133,000 human subjects in a three-year period; eighty-five reported at least one injury among their subjects. In all, 5,000 injuries were noted, of which 4,000 were considered trivial and about 1,000 temporarily disabling. Fifty-seven injuries resulted in death or permanent disability, but these were not necessarily clearly related to the research.

Dr. Philippe Cardon, the main author of the study, concluded that the risks of participation in non-therapeutic research are probably no greater than those in everyday life, and in therapeutic research, no greater than those in other treatment settings [3].

Dr. John Arnold, director of the Quincy Research Center, in Kansas City which engages subjects, mainly for drug studies, similarly reported a small number of injuries at various trials or phases of investigation. However, he reported that many complaints among Quincy volunteers were related to anxiety or depression following the use of potent anti-psychotic medications. As a result, one subject left the Center against medical advice on a motorcycle. Another later fell asleep in bed with a cigarette burning in his hand, and one expressed generalized anger without understanding the reason. A bizarre problem arose when a participant was killed by another, evidently during a robbery attempt soon after both were discharged with their pay. These incidents are most unusual and, obviously, are not related to the drug or procedure but must be reckoned as somehow related to the research experience [2].

Although the number of injuries and associated problems is admittedly small, the concern is great. There is almost unqualified consensus in the research field that research subjects, and perhaps others who incur injury or substantial adversity through association, should be given special consideration. The study on compensation of research subjects undertaken for the Secretary of HEW therefore recommended recovery, to be based on a no-fault principle [7].

In 1978, the Department of HEW (now HHS) issued a regulation requiring research organizations to tell subjects "whether compensation and medical treatment is available if physical injury occurs and, if so, what it consists of or where further information may be obtained."

In 1981, this provision, which is part of the regulation requiring prior informed consent of research participants, was modified to cover research involving *more than minimal risk* but applicable to *injury* of any type, not just physical impairment. This change recognized that most minor injuries were directly, informally treated and yet also recognized injuries that go beyond immediate physical damage (45 *CFR* 46, Sec. 46. 116 (b). The 1981 regulations also require institutions to report unanticipated problems involving risks and also injuries directly to the Government, not solely to local review committees.

PRESENT METHODS

The regulation, it will be noted, does not require any compensation or remedy but only that the subjects be advised. Moreover, the regulation does not apply to private research but only to the extent of Federal control. It was expected that the notice requirement would encourage model programs but, so far, no marked changes have been developed in sponsored or private auspices.

At this time, compensation for injuries is managed in many different ways and, since there are few incidents of serious disability or death, most are handled informally through medical care, payment of medical expenses and reimbursement for minimal economic loss. Research grants often include some allowance for this purpose and many hospitals either cover the bill or pass some or all of it on to third parties. Uniquely, the Quincy Research Center covers its subjects under the state worker compensation law.

It is likely that claims will increase as human research becomes more formalized through the review systems and institutional boards which have now been established, and patients and volunteers recognize that they are reasonably entitled to recompense, particularly for an injury which goes beyond any informed consent expectations or arises through negligence, accident, or mishap.

A few organizations, such as the Quincy Research Center and the University of Washington, have formal systems covered by private or self-insurance and many drug companies cover the risk under their general liability programs. There is no public system, either state or federal, prescribing payments or proposing standards.

INACTION AND OPPOSITION

Inaction, indeed, opposition to any special injury-compensation program, is all too evident. This, despite positive recommendations in 1977 by the HEW Task Force, and by the National Commission for the Protection of Human Subjects (replaced by the President's Commission in 1978) and, later by the President's Commission, which favored some action but suggested an extensive pilot study to review problems, options, and potentials. As early as the 1960's, this author proposed a no-fault approach, based on voluntary contract [9]. At the *Daedalus* and the New York Academy of Sciences conferences on human experimentation, among others, this view was again advanced and amplified [10].

Several reasons are cited overtly, or are manifest by failure or refusal to move:

1. There is no problem! That is, quantitatively, the number and severity of injuries do not warrant any special consideration. The informal arrangements and largess or *ex gratia* payments appear to be sufficient. And, many injuries which are "managed" without formal claim would, under a system, stimulate claims, many without merit, but requiring response and expense.
2. There is no societal or ethical imperative. Arguments by Dr. H. Tristram Engelhardt that society has a debt to volunteers who promote the common good and that the research enterprise has such an obligation in equity, fairness, and as a reasonable business expense are not convincing in light of possible benefits to subjects, especially in therapeutic studies and under legal agreements to take part in research. For example, Alexander Capron, attorney and former Executive Director of the President's Commission states that subjects may validly waive recompense as part of consent arrangements – except for negligence or intentional harm. (He does recognize, however, that such action would sorely try public expectation and policy). In short, there is no intrinsic right or the need to create one [14].
3. There are adequate means for redress under tort (malpractice) liability theories for negligent action, substandard research, or lack of consent. Present insurance arrangements can generally include such coverage and accommodate claims negotiations and settlements.

4. A special program would not necessarily improve research, reduce risks, or prevent injuries, as claimed, since there are already professional motivations and sanctions. Rather, the burdensome administrative aspects of a special program might hinder research.
5. Finally, perhaps most importantly, the problems of determining when and whether the research caused the injury, particularly in therapeutic studies, and the extent of injury are so complex that no formula or system can be devised as fair and feasible. In part, because of this and other uncertainties, insurance for this purpose is not easily available. There would be no financial assurance through conventional, private sources. Government programs would be costly, bureaucratic, and inflexible and, in view of the small number of reported injuries, unnecessary.

Other criticisms have been leveled at the notion of no-fault by some investigators and their sponsors who believe that, without some showing of dereliction, there is no reason for payment. The subject should share the risk as well as the benefit. And, if a normal volunteer, there is usually some payment or fee and, of course, the right to sue in tort or contract.

Many of these arguments are expressed or implied in an essay "Compensation and Cancer Research" which appeared in late 1981 in *The New England Journal of Medicine*. From the viewpoint of therapeutic research on cancer, the authors state that a special program "lacks a clear moral rationale and poses serious practical difficulties" [1].

THE CASE FOR SPECIAL TREATMENT

The staff report of the President's Commission refers to the dichotomy of need on the basis of numbers versus need as a "socially and morally significant issue," apart from magnitude, but because of what it says about our values and attitudes toward the role of human subjects.

Quite simply, harm or injury can always serve as the basis for a claim on some legal theory based on personal injury, contract, or product or service liability. In the absence of a specified or accepted alternative, some or all disputed issues which are not informally settled are generally litigated. In fact, there have been some cases against research investigators, sponsor institutions, public agencies and, principally, drug firms based squarely on a demand for recovery because of asserted negligence or some type of implied liability. Without the clamor of numbers seeking redress and with the availability of the American court system, is there a need for exceptional or special treatment? Moreover, since these individuals, in most cases, have presumably assumed the risks and possible consequences of research, as evidenced by signed consent forms and approvals by Institutional Review Boards, what else could be needed? If anything, it might be argued that research subjects are better protected, safer and more likely to be immediately helped, at least physically, than any other victims of medical misadventure or psychological error. For example, under current federal regulations, subjects involved in federally conducted, sponsored or supported research are entitled to short-term medical care for research-caused injury.

The answer, therefore, must be found not in the law of contract nor the largess of investigator, sponsor or the availability of medical service. It must be seen as a recognized societal or ethical obligation which holds that those who promote the public good through beneficial scientific research and are willing to advance knowledge through lending their bodies, minds and energies must be compensated by the public, not for their efforts, but for injury to themselves.

The HEW Task Force essentially adopted this view, crediting Dr. Tristram Engelhardt with developing its ethical foundation. It regarded compensatory justice as the operative principle, i.e., that form of justice which seeks to redress injury even when there may be no associated fault or blame. It further agreed that consent does not, in and of itself, constitute a waiver of future compensation. And, after struggling with the issues of therapeutic versus non-therapeutic participation, the Task Force concluded that both should be eligible for compensation. The Task Force achieved this resolution by this formula: that subjects should be compensated if "(1) the injury is proximately caused by such research and (2) the injury on balance exceeds that reasonably associated with such illness from which the subject may be suffering, as well as with treatment usually associated with such illness at the time the subject began participation in the research."

Thus, injury to a normal or well volunteer (non-therapeutic) would simply be assessed against a healthy state. For the ill patient, however, as critically argued in the paper on cancer research, it would be most difficult to arrive at cause and then at the measure of difference between disease and research-induced conditions. This "on balance" concept has received a mixed reception, even from those supporting the case for special compensation.

Perhaps the most acceptable arguments rely on a fair and just societal obligation, certainty, and simplicity. These assert that (1) subjecting participants to the rigors of litigation and requiring adverse or even hostile positions from those who share in research efforts is unsuitable and unseemly; (2) establishing a system in advance of injury is appropriate and necessary to a relationship of trust and cooperation, and (3) reparation should depend on the relationship of the parties and on the causal connection of the harm or loss to the clinical study, without regard to fault. "Simply put, an application of the workmen's compensation concept, rather than employer liability or malpractice, would seem feasible" [9].

METHODS

The problems are not unique to the United States nor is the concern for developing fair and satisfactory solutions. England and Sweden, particularly, have given recent attention to providing injury compensation. A British Commission has recommended a no-fault system modeled on a worker compensation plan which would provide a fund for payment. Final approval will depend on the position taken by Common Market countries. Sweden's proposed legislation stimulated a voluntary consortium of major insurers to provide compensation, similarly based on administrative determinations of liability without fault and a scale of recompense. Without law, there is still assurance of reparation.

The survey of eleven other countries prepared for the Task Force disclosed "the almost universal lack of any mechanism dealing specifically with this problem" [5]. Where there is some provision, it is usually associated with some larger scheme of social welfare, national health insurance, or workmen's compensation. The lack of a specific plan, however, does not mean that there is no protection; that depends on the existence of other public or social insurance. Thus, Israel, a typical "no-law" country would, through its Health Ministry, arrange for medical compensation by the sponsoring institution.

For the United States, several approaches have been suggested, essentially either governmentally administered or guaranteed, or privately mandated or adopted, pursuant to regulations or guidelines. In general, these are based on a no-fault principle, similar to or part of a worker-compensation plan, for example, extending coverage under the Federal Employees Compensation Act, or setting up a Federal compensation fund. The no-fault concept can be expressed as a form of strict liability, thus permitting tort suit without requiring proof of negligence and with certain defenses barred, or as a form of entitlement to prescribed benefits payable from an administered fund, on proof of relationship and injury. At the other extreme, investigators or sponsors would be permitted to create approved programs met by private insurance, with or without some Government backstop for contingencies or excess coverage. Also, individual or group disability or health insurance for subjects has been suggested, much like travel insurance for the duration of a project. These would be paid for by the sponsor or investigator; or there could be group insurance for an institution or agency available for subjects as they enter research projects.

ENTITLEMENT PRINCIPLE AND PLAN

The understandable fear of an imposed bureaucracy for an admittedly questionable problem has magnified the issues of definitions, design, and application. First, no simple system need be mandated. Rather, sponsors would be requested to establish, under regulation and guidance, any scheme which is (a) responsive to individual need and societal expectations in providing fair reparation in a dignified way; (b) ethically and legally acceptable with full scope for personal rights and respect for the consent agreement; (c) generally applicable for all studies and subjects; (d) financially stable, and (e) capable of fitting into the research review process [12].

It is likely that some years will elapse before a general plan or federal fund or common organization will be established in the U.S., or in Britain or elsewhere. In the meantime, there will be manifest inconsistency, even incongruity, in the management of a problem which clearly demands a basic policy and reasonably uniform application.

This would not require an identical system for each research sponsor or agency, but at minimum it would call for: (a) a formal plan which would be known to investigators and their patients and to the research community; (b) a basis for payment, preferably through a no-fault arrangement, which could be provided through individual coverage or insurance pool, such as for all universities or hospitals in a given area;

(c) levels of compensation, for example, not less than that provided under local compensation law or equivalent payments under court judgments; and (d) a mechanism for reporting toward the development of more general and uniform systems based on patterns of experience. And there are relatively simple methods, through arbitration or similar processes, for fair, prompt determination of disputed issues [2].

Under such a system, entitlement could be relatively easily defined under conditions such as these:

1. *Automatic entitlement* for recovery (compensation, medical care, rehabilitation) arises when the injury is recognized as one or more of the contingencies or risks set forth (a) in the research protocol or (b) in the consent form or the memorandum of discussion or between subject and research staff, or (c) acknowledged by the investigator as due to participation or for any reason directly related to the research procedure, setting or similar circumstances or (d) acknowledged by the local review committee, sponsor, or insurer group or (e) specified or presumed under any pertinent statute or regulation governing such activity or relationship, e.g., worker compensation law.
2. *Possible entitlement* for recovery arises when the injury is not recognized as one of the contingencies or risks specified above (1(a) through (e)) because of the subject's prior medical or psychological condition. Cases in this group shall be afforded immediate medical care and nominal payment in advance, subject to review.
3. *No entitlement* for recovery arises when the injury is similarly not recognized because of deliberate, willful action by the subject beyond the scope of the research and unrelated to its performance or because of circumstances outside the research context.

In cases of *possible entitlement* or *no entitlement*, any proposals or claims not settled by accord of the parties shall be resolved by arbitration or litigation.

Such a scheme would reasonably classify the claimed injuries and would not pit subject against investigator. In fact, since both have mutual interests, it would be reasonable that they might agree on the causal relationship and entitlement for care and compensation from a fund established for this purpose, as in worker compensation and similar insurance plans.

Disagreements would be resolved amicably through any form of private dispute mechanism, voluntarily accepted. An extensive series of arbitral options was prepared by this author for the President's Commission [13]. Those who did not elect binding arbitration, for example, would be free to sue on negligence or another basis.

The exercise of creative choices and reporting of experience would in due course suggest which approaches are suitable and which issues call for further clarification. It may well be that no governmental system in the United States or elsewhere is needed. As in Sweden, voluntary cooperation of major interests working toward a public interest may prevail. Such formal arrangements would not remove or reduce the advantages of informal practices of immediate care, hospitalization, and payment of out-of-pocket expenses. Structure should not destroy spirit.

There are immediate steps that can and should be taken now by all concerned parties, with good will, for possible later incorporation into an overall system. Thus, may we "honor society's ethical obligation to those who have been subjected to risks in order that research beneficial to society may be undertaken" [6].

Mt. Sinai School of Medicine
New York, New York, U.S.A.

BIBLIOGRAPHY

1. Ackerman, T. F., and Mauer, A. M.: 1981, 'Compensation and Cancer Research', *New England Journal of Medicine* **305**, 760–763.
2. Arnold, J. D.: 1980, 'Incidence of Injury During Pharmacologic Research and Indemnification of Injured Research Subjects at the Quincy Research Center', *Report* submitted to the President's Commission.
3. Cardon, P. W., Dommel, J. and Trimble, R. R.: 1976, 'Injuries to Research Subjects: A Survey of Investigators', *New England Journal of Medicine* **295**, 650–654.
4. Engelhardt, H. T.: 1977, 'Study of the Federal Government's Ethical Obligations to Provide Compensation for Persons Injured in the Course of Their Participation in Research Supported by Funds Administered by the Secretary of H.E.W.', in HEW *Task Force on the Compensation of Injured Research Subjects*, Appendix A, 45–64.
5. Haubenreich, J. G.: 1977, 'Compensation of Injured Research Subjects – A Comparative Analysis', in HEW *Task Force on the Compensation of Injured Research Subjects*, Appendix A, 65–79.
6. Havighurst, C. C.: 1977, 'Mechanisms for Compensating Persons in Human Experimentation', in HEW *Task Force on the Compensation of Injured Research Subjects*, Appendix A, 81–132.
7. HEW: 1977, HEW *Secretary's Task Force on the Compensation of Injured Research Subjects*, (Jan.), VIII–IX.
8. *HHS News*: (Jan. 20) 1982, (FDA) Release No. P82–3.

9. Ladimer, I.: 1963, 'Clinical Research Insurance', *Journal of Chronic Diseases* **16**, 1229–1235.
10. Ladimer, I.: 1969, 'Protection and Compensation for Injury In Human Studies' in Freund, P. A. (ed.) *Experimentation with Human Subjects*, George A. Braziller, New York, 247–261.
11. Ladimer, I.: 1970, [Opening Remarks] 'New Dimensions in Legal and Ethical Concepts for Human Research', *Annals of the New York Academy of Sciences* **169**, 297–298.
12. Ladimer, I.: 1977, 'Positive Protection: A Proposal for Compensation and Other Remedies for Subjects Injured in Biomedical Research', in HEW *Task Force on the Compensation of Injured Research Subjects*, Appendix A, 133–146.
13. Ladimer, I.: 1980, 'Arbitral Processes for a Program to Compensate Injured Research Subjects', *Report* submitted to the President's Commission.
14. O'Donnell, T. J.: 1981, 'Reply to Irving Ladimer', in *Compensation for Research Injuries: The Medical-Moral Newsletter* **18** (Nos. 2, 3, 4).
15. President's Commission for the Study of Ethical Problems in Medicine and Biomedical and Behavioral Research: (Dec.) 1981, *Protecting Human Subjects*, U.S. Government Printing Office, Washington, D.C.

APPENDIX

ISRAEL HEALTH REGULATIONS: EXPERIMENTS ON HUMAN SUBJECTS – 1980

ISRAEL HEALTH REGULATIONS: EXPERIMENTS ON HUMAN SUBJECTS (1980)

In my capacity, as specified in section 33 of The National Health Regulations, 1940, I (The Director General of the Ministry of Health of Israel) hereby issue the following regulations:

1. DEFINITIONS:

In these regulations
"Accredited Medical School" is any medical school accredited by the Council on Higher Education.

"The Directory" is the directory of drugs as defined in the Pharmacists Regulations (Medical Index, 1977, henceforth the Pharmacists Regulations).

"Helsinki Declaration" is the declaration laying down the recommended guidelines for physicians carrying out experiments on human subjects (*Viz.*, Helsinki, 1974, as revised in Tokyo 1975, which version is given in the Appendix).

"Helsinki Committee" is the Committee according to section 2 chapter one of the Helsinki Declaration.

"Hospital Helsinki Committee" is the Committee whose legal composition, appointment, and number are as stated in Appendix Two, and whose function is to authorize every medical experiment on human subjects carried out in hospital.

"Supreme Committee" is the Supreme Helsinki Committee for medical experiments on human subjects, whose legal composition, appointment, and number are specified in Appendix Three, appointed for general or for specific purpose, and whose function is to issue an opinion in every matter covered in Regulation 3(2).

"FDA" is the Food and Drug Administration in the Ministry of Health.

"Medical Experiment on Human Subjects" (1) refers to the use of materials, in contravention to the legal authorized use of that material, or when the standard use of the material in Israel today is not that which is being requested, or has not been tested in Israel, and has potential or being designed to affect the physical or mental health of the person, or the embryo, or part of them, including the genetic system; (2) includes any procedure, act or examination on a human that is not standard practice.

"The Director" is the Director General of the Ministry of Health, or the person ordained by him for the purpose of these regulations, in total or in part.

AUTHORIZATION FOR EXPERIMENT:

2. (a) No experiment shall be performed on a human in hospital without the written authorization of the Director, and shall be in accordance with the conditions specified in that authorization.
(b) No experiment will be performed on a human in hospital in contravention to these regulations and the Helsinki Declaration.

CONDITIONS FOR AUTHORIZATION OF EXPERIMENTS:

3. The Director shall not issue an authorization for medical experiments on human subjects unless:

(1) the Helsinki Committee of the hospital where the experiment will take place informs the Director, in writing, of its authorization for the experiment (henceforth called the Notice of Authorization for an Experiment);

(2) the Director is convinced that the experiment is not in contravention to the Helsinki Declaration and these regulations;

(3) a judgement by the FDA or the Supreme Committee has been accepted in matters covered in 3a and 3b (see below) respectively; in this section "a judgement by the FDA" denotes any of its committees as appointed by the Director.

EXPERIMENTS REQUIRING AUTHORIZATION OF THE FDA:

3a. The following experiments on human subjects may be authorized by the Director only after obtaining the opinion of the FDA:

(1) experiments on human subjects with drugs not listed in the Directory;

(2) experiments with drugs listed in the Directory, but not in accordance with the conditions and authorized use previously granted these drugs.

EXPERIMENTS REQUIRING AUTHORIZATION OF THE SUPREME COMMITTEE:

3b. The following experiments on human subjects may be authorized by the Director only after obtaining the opinion of the Supreme Committee:

(1) experiments affecting the genetic system of man;

(2) experiments affecting the female fertility in an unnatural [artificial] manner;

(3) any other matter that the Director requests in order to determine whether regulation 3(2) applies.

SUBMISSION OF THE REQUEST:

4. A request for authorization for a medical experiment on human subjects must be submitted, in writing, to the Director by the director of the hospital in which the experiment will be carried out and by the physician responsible for the experiment; a detailed protocol for the requested experiment must be attached, which includes the purpose of and need for the experiment, and details of similar research and experimental tests already performed in Israel or abroad.

ADDITIONAL INFORMATION:

5. The Director is empowered to request additional details on every aspect of the requested experiment, at any time, both before and after authorization is issued.

ADDITIONAL INFORMATION AND CANCELLATION OF AUTHORIZATION:

6. The Director is empowered at any time to alter the conditions of the authorization, to limit it, to add conditions, and to cancel it.

EVIDENCE:

7. Anyone claiming receipt of authorization according to these regulations must provide evidence of that authorization.

AUTHORIZATION FOR MEDICAL RESEARCH:

8. Authorization of the Director for medical or scientific research according to regulation 17(5) of the Pharmacists Regulations (Medical Index, 1977), is considered as authorization given according to these regulations.

PRESERVATION OF LAW:

9. These Regulations are in addition to legal judgement and codes of ethics; no right, confirmation or authorization granted by these Regulations, or in accordance with them, shall supersede the obligations imposed in any legal judgement or codes of ethics.

EFFECT:

10. These regulations shall take effect 45 days from their publication.

APPENDIX I (REGULATION I)

DECLARATION OF HELSINKI

APPENDIX II (REGULATION I)

HOSPITAL HELSINKI COMMITTEE

APPENDIX II (REGULATION I)

THE SUPREME (HELSINKI) COMMITTEE

EDITORS' NOTE

The Health Regulations which govern, in Israel, the use of human beings in research are promulgated under the authority of The Director General of Israel's Ministry of Health. Only Appendix I (Regulations I), Declaration of Helsinki, is reprinted here, since it is the principal governing document germane to research on human subjects in Israel. All prior corrections to the original Code of Regulations of December

11, 1980 (4189, p. 292), are included and can be found in the following Codes of Regulations in Israel:
For corrections see:

Code of Regulations 4376, July 1, 1982, page 1272;
Code of Regulations 4635, May 24, 1984, page 1570;
Code of Regulations 4706, September 24, 1984, page 2646;
Code of Regulations 4733, November 23, 1984, page 367.

The pertinent Health Regulations were translated into English from the Hebrew by Professor Avner Cohen and his colleagues of the Department of Philosophy, Tel Aviv University, Ramat Aviv, Israel. The Editors are extremely grateful to the translators for making the English translation of the authorizing document available for the first time.

APPENDIX I

(REGULATION I)

DECLARATION OF HELSINKI

Recommendations guiding medical doctors in biomedical research involving human subjects

Adopted by the 18th World Medical Assembly, Helsinki, Finland, 1964 and As Revised by the 29th World Medical Assembly, Tokyo, Japan, 1975.

INTRODUCTION

It is the mission of the medical doctor to safeguard the health of the people. His or her knowledge and conscience are dedicated to the fulfillment of this mission.

The Declaration of Geneva of the World Medical Association binds the doctors with the words, "The health of my patient will be my first consideration", and the International Code of Medical Ethics declares that, "Any act or advice which could weaken physical or mental resistance of a human being may be used only in his interest."

The purpose of biomedical research involving human subjects must be to improve diagnostic, therapeutic and prophylactic procedures and the understanding of the aetiology and pathogenesis of disease.

In current medical practice most diagnostic, therapeutic or prophylactic procedures involve hazards. This applies *a fortiori* to biomedical research.

Medical progress is based on research which ultimately must rest in part on experimentation involving human subjects.

In the field of biomedical research a fundamental distinction must be recognized between medical research in which the aim is essentially diagnostic or therapeutic for a patient, and medical research, the essential object of which is purely scientific and without direct diagnostic or therapeutic value to the person subjected to the research.

Special caution must be exercised in the conduct of research which may affect the environment, and the welfare of animals used for research must be respected.

Because it is essential that the results of laboratory experiments be applied to human beings to further scientific knowledge and to help suffering humanity. The World Medical Association has prepared the following recommendations as a guide to every doctor in biomedical research involving human subjects. They should be kept under review in the future. It must be stressed that the standards as drafted are only a guide to physicians all over the world. Doctors are not relieved from criminal, civil and ethical responsibilities under the laws of their own countries.

I. BASIC PRINCIPLES

1. Biomedical research involving human subjects must conform to generally accepted scientific principles and should be based on adequately performed laboratory and animal experimentation and on a thorough knowledge of the scientific literature.

2. The design and performance of each experimental procedure involving human subjects should be clearly formulated in an experimental protocol which should be transmitted to a specially appointed independent committee for consideration, comment and guidance.

3. Biomedical research involving human subjects should be conducted only by scientifically qualified persons and under the supervision of a clinically competent medical person. The responsibility for the human subject must always rest with a medically qualified person and never rest on the subject of the research, even though the subject has given his or her consent.

4. Biomedical research involving human subjects cannot legitimately

be carried out unless the importance of the objective is in proportion to the inherent risk to the subject.

5. Every biomedical research project involving human subjects should be preceded by careful assessment of predictable risks in comparison with foreseeable benefits to the subject or to others. Concern for the interests of the subject must always prevail over the interest of science and society.

6. The right of the research subject to safeguard his or her integrity must always be respected. Every precaution should be taken to respect the privacy of the subject and to minimize the impact of the study on the subject's physical and mental integrity and on the personality of the subject.

7. Doctors should abstain from engaging in research projects involving human subjects unless they are satisfied that the hazards involved are believed to be predictable. Doctors should cease any investigation if the hazards are found to outweigh the potential benefits.

8. In publication of the results of his or her research, the doctor is obliged to preserve the accuracy of the results. Reports of experimentation not in accordance with the principles laid down in this Declaration should not be accepted for publication.

9. In any research on human beings, each potential subject must be adequately informed of the aims, methods, anticipated benefits and potential hazards of the study and the discomfort it may entail. He or she should be informed that he or she is at liberty to abstain from participation in the study and that he or she is free to withdraw his or her consent to participation at any time. The doctor should then obtain the subject's freely-given informed consent, preferably in writing.

10. When obtaining informed consent for the research project the doctor should be particularly cautious if the subject is in a dependent relationship to him or her and may consent under duress. In that case the informed consent should be obtained by a doctor who is not engaged in the investigation and who is completely independent of this official relationship.

11. In case of legal incompetence, informed consent should be obtained from the legal guardian in accordance with national legislation. Where physical or mental incapacity makes it impossible to obtain informed consent, or when the subject is a minor, permission from the responsible relative replaces that of the subject in accordance with national legislation.

12. The research protocol should always contain a statement of the ethical considerations involved and should indicate that the principles enunciated in the present Declaration are complied with.

II. MEDICAL RESEARCH COMBINED WITH PROFESSIONAL CARE

(Clinical Research)

1. In the treatment of the sick person, the doctor must be free to use a new diagnostic and therapeutic measure, if in his or her judgment it offers hope of saving life, reestablishing health or alleviating suffering.

2. The potential benefits, hazards, and discomfort of a new method should be weighed against the advantages of the best current diagnostic and therapeutic methods.

3. In any medical study, every patient – including those of a control group, if any – should be assured of the best proven diagnostic and therapeutic methods.

4. The refusal of the patient to participate in a study must never interfere with the doctor-patient relationship.

5. If the doctor considers it essential not to obtain informed consent, the specific reasons for this proposal should be stated in the experimental protocol for transmission to the independent committee (I, 2).

6. The doctor can combine medical research with professional care, the objective being the acquisition of new medical knowledge, only to the extent that medical research is justified by its potential diagnostic or therapeutic value for the patient.

III. NON-THERAPEUTIC BIOMEDICAL RESEARCH INVOLVING HUMAN SUBJECTS

(Non-clinical biomedical research)

1. In the purely scientific application of medical research carried out on a human being, it is the duty of the doctor to remain the protector of the life and health of that person on whom biomedical research is being carried out.

2. The subjects should be volunteers – either healthy persons or patients for whom the experimental design is not related to the patient's illness.

3. The investigator or the investigating team should discontinue the research if in his/her or their judgment it may, if continued, be harmful to the individual.

4. In research on man, the interest of science and society should never take precedence over considerations related to the well-being of the subject.

NOTES ON CONTRIBUTORS

Ilai Alon, Ph.D., is Senior Lecturer, Faculty of Humanities, Tel Aviv University, Ramat-Aviv, Israel.

William F. Bynum, Ph.D., is Head, Unit of the History of Medicine, Department of Anatomy and Embryology, University College, and Assistant Director of Research, Wellcome Institute for the History of Medicine, London, England.

Arthur L. Caplan, Ph.D., is Director, Center for Biomedical Ethics, University of Minnesota, Minneapolis, Minnesota.

Linda Rabinowitz Dagi, M.D., is a Resident in Opthalmology, Center for Sight, Georgetown University Hospital, Washington, D.C.

T. Forcht Dagi, M.D., M.P.H., is a member of the Neurosurgery Service, Walter Reed Army Medical Center and Senior Research Fellow, Kennedy Institute of Ethics, and Adjunct Professor of Law, Georgetown University, Washington, D.C.

Corinna Delkeskamp-Hayes, Ph.D., Buchbergstrasse 17, D-6463 Freigericht 1, Federal Republic of Germany.

H. Tristram Engelhardt, Jr., Ph.D., M.D., is Professor of Medicine and of Community Medicine, and Member, Center for Ethics, Medicine and Public Issues, The Baylor College of Medicine, Houston, Texas.

Anne Fagot-Largeault, Ph.D. (Docteur de lettres), M.D., est professeur de philosophie à l'Université de Parix-X. Nanterre; et médecin attaché (psychiatrie), à l'hôpital Henri Mondor, Créteil, France.

Irving Ladimer, S.J.D., is Adjunct Professor of Community Medicine, Mt. Sinai School of Medicine, New York, New York.

Robert U. Massey, M.D., is Professor of Medicine and Community Medicine and Health Care, School of Medicine, University of Connecticut Health Center, Farmington, Connecticut.

Michael Ruse, Ph.D., is Professor of History and Philosophy, Departments of History and Philosophy, University of Guelph, Ontario, Canada.

Hans-Martin Sass, Ph.D., is Senior Research Fellow, The Kennedy Institute of Ethics, Georgetown University, Washington, D.C., and Professor of Philosophy, Institut für Philosophie, Ruhr-Universität Bochum, Bochum-Querenburg, West Germany.

Stanley G. Schade, M.D., is Professor of Medicine, University of Illinois at Chicago, Illinois.

Amos Shapira, J. D., is Professor of Law, The Faculty of Law (The Kalman Lubowski Chair in Law and Biomedical Ethics), Tel Aviv University, Ramat-Aviv, Israel.

Stuart F. Spicker, Ph.D., is Professor of Community Medicine and Health Care, School of Medicine, University of Connecticut Health Center, Farmington, Connecticut.

Kenneth L. Vaux, Ph.D., is Professor of Ethics, University of Illinois at Chicago, Illinois.

Andre de Vries, Ph.D., M.D., is Emeritus Professor of Medicine, Tel Aviv University, Ramat-Aviv, Israel.

INDEX

Ackernecht, E. H. 31, 36
acne 12
adverse reactions 212
AIDS xv, 5, 217
Aiken, John 36
Altman, Douglas 175–177
Altman, Lawrence K. 249–250, 255
altruistic case 14
American Medical Association 53, 67
American Psychological Association 53
American Society for Clinical
    Investigation 41
Amundsen, Darrel 125
Aquinas, St. Thomas 7
Aristotle 6, 154, 165–166
Arnold, John 262
Arzneimittelgesetz (AMG) 56–60,
    63–64, 70
asthma 212
aurothiomalate 212
auto-experimentation 249–254, 256–259
auto-observation 249
    see auto experimentation
    see self-examination
    see National Institutes of Health
autopsy 164–165
Avicenna (Arabic physician) 147
azidothimodine (AZT) 5

B-blocking agent 212
Bacon, Francis 127–129
Barber, Bernard 94, 232
Bayle, 37
Bean, William B. 249
beclomethasone 212
Beecher, Henry K. 3, 8, 10, 161, 230, 250
Belmont Report 4, 132, 167
beneficence 132, 134, 136, 137, 256
Bernard, Claude 30, 39, 146, 147, 161, 165
Bert, Paul 39

Beveridge, William 22
Bichat, Xavier 37, 129
bioengineering technology 145
bioethics 26, 229, 233
biomedical research 48, 98
    see human experimentation
blind/double blind 132–133, 135, 179
blood transfusions 35
Bok, Sissla 91
Bonet, Theopile 129
Brandt, Karl 6, 230
Branson, Roy 68
British Medical Research Council 53, 190
Broussai, 37
Brown v. Hughes 126
Browne, Sir Thomas 20

Cajetan 125–126
cancer 12, 211
Capron, Alexander M. 264
carcinogens 211
Caron, Philippe 262
Carpenter v. Blake 126
Cavers, D.C. 61
Celsus 33
Chaimis, Batholomaeus 125
Chain 40
Chalmers, Thomas C. xvii, 140,
    148–149, 151–155
Cheselden, William 35
children, use of in research 68, 69
    issue of recruiting 104
    rights of protected 105
    reference to (Israel) 112
cholera 250
clioquinol 212
clozapine 212
Code of Deontology (Paris) 162
Codes: legal, medical, research,
    organizations 52–55
Cohen, Avner 278

compensation xii, xv, xx, 66–67, 70–71, 73
consent rule 91
    see informed consent
contrimoxazole 212
controlled clinical test 60
    blind/double blind 5, 132–133, 135, 156, 179
    early development of 128–130
    metaphorically 207
    pilot 147, 150–154
    randomized xvi, xvii, 132, 135, 147–155, 179, 181, 192
    statistics in 213, 223–224
    workshop, (Italy) 62–63
    see randomized clinical trial
Copernicus, Nicholas 145
Corvisart, 37
cost/benefit calculation 201
coumadin 5
Council of American Physiological Society 112
Council for International Organizations of Medical Sciences (CIOMS) 54, 71–72
cromoglycate 212
Cullen, William 129
Curran, William 126
Cushny 42

Daedalus 264
Dagouman Report (1982) 166, 185
Darwin, Charles 24
demercaprol 202
Department of Health, Education, and Welfare
    Task Force 261
    regulation 262, 264
Department of Health and Human Services 145, 162, 261
depressive patients, depression 177
deregulation (forms of) 74
deRubeis, R. J. 177
digoxin 5
disease
    AIDS 5, 217

Hodgkins' Disease 9
leukemia 9
scabies 30
scurvy 35
sickle-cell anemia 9
small pox 35
syphilis 35

Earle, Pliny 130
Einstein, Albert 5
Eisenberg, Leon 234
Emmerich 249
endorsement rule 94
    endorsibility xv
Engelhardt, Jr., H. Tristram 5, 143–144, 264
Enlightenment 8, 23
Entralgo, P. Laín 32
Erasistratus 33
Ethical Committees 50, 53
ethics
    experimentation 165
    informed consent 133–134
    necessity 174
    of beneficence 161
    of knowledge 161, 166–167
    pilot trials 147
    RCT 154–156, 163
    research 163, 180, 182
    science 166
    scientific 165, 177
    standards 163
    tension in 145–146
    unethical 175–176
experiment
    defined 255
experimental medicine 39
    branches of 39
    pathology 39
    pharmacology 39
    statistics 38
experimentum periculosum 21

fair play 236–242, 244–245
Fincke, Martin
    views testing drugs 61–62

First International Congress of Medical
    Ethics 163
Fisher, R. A. 133
Florey, Howard 21, 40
Flexner, Abraham 41
Food and Drug Administration (FDA)
    51, 54, 66, 162, 211–212, 260,
    275–276
  see United States
Forsmann, W. 249
Fose, Rene 94
Foucault, Michel 36, 129
Freidson, Eliot 94, 96
Freiman, J. E. 158
Freud, Sigmund 124, 257
Freund, Paul A. 69, 70, 230
Fried, Charles 91, 234
Fumus, Batholomaeus 125

Galen, 21, 33
Galilei, Galileo 144–146
Galton, Francis 38
Gavaret, J. 130
Gaylin, William 27
Gelfand 36
Geneva formulations 4
German Reich 51
Giere, R. 192
Golden Rule 252, 256
Gordon, Bernard de 34
Gorovitz, Samuel 27
Gray, Bradford 91
Guidelines for the Protection of Health
    and Human Services 4
guinea pigs 29–30, 34, 208
  see human experimentation

Hagans, J. S. 153
Hallerworden-Spatz Syndrome 164
Hart, H. L. A. 236, 238
Harvey, McGehee 41
Harvey, William 21
Head, Sir Henry 40
Hegel, G. W. F. 94–100
Helsinki Declaration 51, 110–114,
    275–276

Risk/Benefit calculus 111
informed consent doctrine 111
Sheebs Medical Center, guided by
    112
Helsinki formulations 4
  Helsinki Committees xvi, 276
hepatitis 230
Herophilus 33
herpes 217
Hill, Austin Bradford 189, 208–209
Hippocratic Corpus xiv, 31
  Oath 65, 76, 162
Hollon, S. D. 177
Hopkins, Johns 41
Horsley, Sir Victor 41
human experimentation xiv, 30, 36, 38,
    50, 54, 74, 76, 126, 131, 135–136,
    230–232, 234–235
  biblical viii
  compensation 262–268
  control models 103–106, 129
  defined 132, 134
  human subjects 94
  issues 103, 164
  medicine's role 130
  occupation 67
  order of risk 236, 255
  plan 268–269
  reciprocity and compensation
      235–236, 263–264
    see insurance
  research 48, 98
  to pursue goals 125–126
    see Belmont Report 132
    see beneficence
    see biomedical research
    see controlled clinical test
    see Department of HEW
      regulations
    see guinea pigs
    see new therapy
    see research
human experimentation (Germany)
  rules of 59
human experimentation (Israel) 110–113
  definitions (Israel) 275–276

disadvantaged people 167, 172, 229, 231–233
  extends to 109
  inequities 233
  informed consent, concepts of 116
  Israel Health Regulations 275, 282
  Jewish law 254
  obligations in research institution 240–241
  obligations in teaching hospitals 239–240
  underlying issues of 114
  see Israel Health Regulations
humanitarianism 24
humanitarian law 6
humoral medicine 31, 33
Hunter, Thomas 27, 35
Huntington's chorea 169, 170
Hutchinson, Jonathan 124
hydralazine 212
hypertension 212

Idealogues 24
informed consent 9, 13, 15, 20, 69, 73, 91, 104, 133–134, 230, 233, 241
  ethics in 133–134, 229, 231
  in research institutions 242
  inequities 236
  Israeli tort 108
  legal-medical guidelines (Israel)
    informed consent forms 112, 266
    early use of 130
  permission 126
  reason for 132
  teaching hospitals 240
  voluntary 91
  see consent rule
Institutional Review Boards (IRBs) 10, 53, 70, 74, 91, 105, 163, 175, 242, 266
  committee review 230
  see peer review committees
insulin 202
insurance 64, 78, 104, 264
  compensation fund 71
  Federal Employees Compensation Act 268

insurance companies 74
workers' compensation insurance xv, 267
  see human experimentation
in vitro 115, 201, 208–209
  see in vivo
in vivo 201
  see in vitro
Israel Health Regulations 275–282
  see human experimentation (Israel)
Ivy, Andrew, 3–8

Jacobovits, I. 253–254
Jenner, Edward 35, 130
Jewish Chronic Diseases Hospital (case) 230
Jonas, Hans 66, 73, 91, 93, 126, 132, 136, 144, 145, 149, 174, 199–200, 213–214, 234–235

Kant, Immanuel 6, 8, 14, 178
  see moral theories
Katz, Jay 117, 126, 230
Koch, Robert 23, 25, 249
Krebiozen, 3, 6
Kupat Cholim 111
  (Health Service of the Israeli Federation)

L-dopa 202
Laennec, R.-T.-H. 37
Lasagna, Louis 67, 155–157, 234
Lewis, Thomas 41
Lind, James 35
Lindley, D. V. 175
Louis, Pierre 37
Louis, P. C. A. 130

Ma'ayan, Reuven (Israel)
  case of 115
  use of DMB6 115
Mackenzik, 42
May, W. W. 176
McCarthy Era 3
McConce, E. A. 255–256
McDermott, Walsh 234
McKeown, T. 40

medical research
  see human experimentation
Medical Research Council 42
Meier, Paul 153, 207–209
Meister, Joseph 23, 130
melanoma 12
Mellanby, Kenneth 29
Meltzer, Samuel James 41
Method of Difference 147, 148, 149
  see Mill, John S.
Mill, John Stuart 147, 215
  see Method of Difference
Monod, Jacques 166, 169
moral theories 206–207
  utilitarian ethics 206
  Kantian 206
    see Kant, I.
    see utilitarian ethics
Morgagni, G. 129
multiple sclerosis 169–170

National Commission for the Protection of Human Subjects of Biomedical and Behavioral Research 67, 132, 167, 171, 264
  see President's Commission
National Eye Institute Workshop 148
National Institutes of Health 53
  see auto-observation
National Science Foundation 24
Naturphilosophie 24
Naval Research Institute 7
Nazi experiments 6
  Nazi 230
  Nazi Germany 10, 126
New England Journal of Medicine 265
new theraphy: defined 50–51, 74, 76
  different from scientific experimentation 50
Newton, Sir Isaac 24
New York Academy of Sciences 264
Nobel Prize 5, 166, 200, 206
non-therapeutic experimentation 51
Nozick, Robert 237
null hypothesis 155, 203, 205
Nuremberg 230
Nuremberg Code xiii, xiv, 3, 8, 51–52, 66, 68, 72, 161, 254
  defined 52
  see Nuremberg Trials
Nuremberg Trials (War Crimes) 3, 6, 14, 19, 29
  see Nuremberg Code

Oath of Hippocrates 14
one-tailed and two-tailed experiments 204
oral contraception 211
Oxford University 21

Paget, Sir James 40
Pappworth, M. H. 30
Paris Court of Appeal 162
Parkinson's disease 202
Pasteur, Louis 23, 130
pathological medicine 37
Pearson, Karl 38
peer review committees 233
  see Institutional Review Boards
penicillamine 212
penicillin 11, 21, 40, 190
Pettenkofer 249
Pharmacists' Regulations (Israel) 113
Phipps, James 37, 130
Pilatus (ethics of) 185
pilot trials 140–141, 142–143, 175
  in auto-experimentation 257
placebo 5, 11, 124, 148, 172, 174, 177–179, 181, 230
Plato 7, 154
Pope Pius XII 161
Popper, Karl 171, 195
practolol 212
President's Commission for the Study of Ethical Problems in Medicine and Biomedical and Behavioral Research 261, 264, 270
  see National Commission
primaguine 212
prisoners of war 65
  criteria for recruiting 104
  prisoners 66–69
  reference to Israel 112
  rights of protected 105

use of, 65
professional ethics 65
professional research subjects 74
   subcontractors 75
psychotherapy 177
Ptolemy 33
Public Health Service 190
   see United States
public policy 65
Purkinje, J. E. 257
Pythagoras 154

Quincy Research Center 262–263

randomized clinical trial (RCT) xvi, xvii
   see controlled clinical test
Rawls, John 206, 236, 238
Reed, Sir Walter 40, 130, 235
regulation xiv, xx, 55, 58, 64–65,
   66, 71–72
Relman, Arnold 5
research 19–21, 22, 24–26, 42
   use of statistics 176–179
   theoretic 178–179
   institutions 245
      see human experimentation
Richtlinien 50–51, 76
risks in research 262
   non-therapeutic 262
Rockefeller Institute 41
   Foundation 42
Roman Church 144–145
Ross, W. D. 253
Royer, Pierre 162–163

Sanctorius 249
Sass, Hans-Martin 92, 94
Schopenhauer, Arthur 174
Schwartz, Daniel 178, 183
Science 21, 23–24, 26, 42
scientific excellence 3, 4, 6
Segizi, Father M. 144
self-examination 254, 261
   see auto-observation
Seneca 7
Shaw, George Bernard 30

Sheeba Medical Center 112
   Guidelines for the Conduct of
      Research
Shryock, Richard 24
Simmons, John A. 237
Simon, Julian 153
social contract 235–236
Southwest Foundation 230
Spallanzani 35
Spinoza, B. 76
sponsors 214–215
statisticians 38–39, 177
   approach 179, 184
streptomycin 207
   Tuberculosis Trials Committtee 208
suxamethonium 212
Sydenham, Thomas 128–129
syphilis 8, 230

technology 24–26
test (of significance) 192
thalidomide 191, 210, 212, 215
Toulmin, Stephen 91
Towers, Bernard 21, 25
triacetyloleandomycan 12
tricyclic antidepressant drugs 177
triparanol 212
tuberculosis 23, 207–208
Tuskegee xiii, 8
   Syphilis Study 126, 230

United States 146
   see Food and Drug Administration
   see Public Health Service
urinary infections 212
utilitarian ethics 206–207
   see moral theories

Vibrio Cholerae 249
Villanova, Arnold of 34
vivisection 33, 165
vulnerable populations 114, 116
   children 68–69
   economically disadvantaged xiv
   exploitative 126
   inequities xix

poor 125
prisoners 68
prisoners of war 69

Wartofsky, Marx 68
Wenckebach, 42
Western medicine 31
Willowbrook Study 230

Wollman, Elie L. 166
World Health Organization (WHO) 54, 61, 71–72
World Medical Association (WMA) 163, 280
Wulff, Henrik R. 130

Zelen, M. 181

The Philosophy and Medicine Book Series

Editors

H. Tristram Engelhardt, Jr. and Stuart F. Spicker

1. **Evaluation and Explanation in the Biomedical Sciences**
   1975, vi + 240 pp. ISBN 90-277-0553-4
2. **Philosophical Dimensions of the Neuro-Medical Sciences**
   1976, vi + 274 pp. ISBN 90-277-0672-7
3. **Philosophical Medical Ethics: Its Nature and Significance**
   1977, vi + 252 pp. ISBN 90-277-0772-3
4. **Mental Health: Philosophical Perspectives**
   1978, xxii + 302 pp. ISBN 90-277-0828-2
5. **Mental Illness: Law and Public Policy**
   1980, xvii + 254 pp. ISBN 90-277-1057-0
6. **Clinical Judgment: A Critical Appraisal**
   1979, xxvi + 278 pp. ISBN 90-277-0952-1
7. **Organism, Medicine, and Metaphysics**
   Essays in Honor of Hans Jonas on his 75th Birthday, May 10, 1978
   1978, xxvii + 330 pp. ISBN 90-277-0823-1
8. **Justice and Health Care**
   1981, xiv + 238 pp. ISBN 90-277-1207-7 (HB)/90-277-1251-4 (PB)
9. **The Law-Medicine Relation: A Philosophical Exploration**
   1981, xxx + 292 pp. ISBN 90-277-1217-4
10. **New Knowledge in the Biomedical Sciences**
    1982, xviii + 244 pp. ISBN 90-277-1319-7
11. **Beneficence and Health Care**
    1982, xvi + 264 pp. ISBN 90-277-1377-4
12. **Responsibility in Health Care**
    1982, xxiii + 285 pp. ISBN 90-277-1417-7
13. **Abortion and the Status of the Fetus**
    1983, xxxii + 349 pp. ISBN 90-277-1493-2
14. **The Clinical Encounter**
    1983, xvi + 309 pp. ISBN 90-277-1593-9
15. **Ethics and Mental Retardation**
    1984, xvi + 254 pp. ISBN 90-277-1630-7
16. **Health, Disease, and Causal Explanations in Medicine**
    1984, xxx + 250 pp. ISBN 90-277-1660-9
17. **Virtue and Medicine**
    Explorations in the Character of Medicine
    1985, xx + 363 pp. ISBN 90-277-1808-3
18. **Medical Ethics in Antiquity**
    Philosophical Perspectives on Abortion and Euthanasia
    1985, xxvi + 242 pp. ISBN 90-277-1825-3 (HB)/90-277-1915-2 (PB)

19. **Ethics and Critical Care Medicine**
    1985, xxii + 236 pp. ISBN 90-277-1820-2
20. **Theology and Bioethics**
    Exploring the Foundations and Frontiers
    1985, xxiv + 314 pp. ISBN 90-277-1857-1
21. **The Price of Health**
    1986, xxx + 280 pp. ISBN 90-277-2285-4
22. **Sexuality and Medicine**
    Volume I: Conceptual Roots
    1987, xxxii + 271 pp. ISBN 90-277-2290-0 (HB)/90-277-2386-9 (PB)
23. **Sexuality and Medicine**
    Volume II: Ethical Viewpoints in Transition
    1987, xxxii + 279 pp. ISBN 1-55608-013-1 (HB)/1-55608-016-6(PB)
24. **Euthanasia and the Newborn**
    Conflicts Regarding Saving Lives
    1987, xxx + 313 pp. ISBN 90-277-3299-4 (HB)/1-55608-039-5(PB)
25. **Ethical Dimensions of Geriatric Care Value Conflicts for the 21st Century**
    1987, xxxiv-298 pp. ISBN 90-55608-027-1
26. **On the Nature of Health**
    An Action-Theoretic Approach
    1987, xviii + 204 pp. ISBN 1-55608-032-8
27. **The Contraceptive Ethos**
    Reproductive Rights and Responsibilities
    1987, xxiv + 254 pp. ISBN 1-55608-035-2
28. **The Use of Human Beings in Research**
    With Special Reference to Clinical Trials
    1988, ISBN 1-55608-043-3
29. **The Physician as Captain of the Ship**
    A Critical Appraisal
    1988, ISBN 1-55608-044-1
30. **Health Care Systems**
    Moral Conflicts in European and American Public Policy
    1988, ISBN 1-55608-045X
31. **Death: Beyond Whole-Brain Criteria**
    1988, ISBN 1-55608-053-0